BAD
HARVEST?

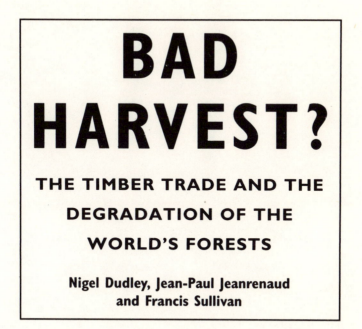

BAD HARVEST?

THE TIMBER TRADE AND THE DEGRADATION OF THE WORLD'S FORESTS

Nigel Dudley, Jean-Paul Jeanrenaud
and Francis Sullivan

WWF

EARTHSCAN
Earthscan Publications Ltd, London

First published in the UK 1995 by
Earthscan Publications Limited

A catalogue record for this book is available from the British Library

ISBN: 1 85383 188 3

Type setting and page design by PCS Mapping & DTP, Newcastle upon Tyne

Printed and bound by Biddles Ltd, Guildford and Kings Lynn

Cover design by

For a full list of publications please contact:
Earthscan Publications Limited
120 Pentonville Road
London N1 9JN
Tel. (0171) 278 0433
Fax: (0171) 278 1142

Earthscan is an editorially independent subsidiary of Kogan Page Limited and publishes in
association with the WWF-UK and the International Institute for Environment and
Development.

Contents

Acknowledgements

A book like this inevitably draws on the expertise and advice of people in many countries. Whilst it is impossible to thank everyone who has contributed to our thinking about the timber trade, we would like to acknowledge in particular the contribution of our colleagues in WWF and elsewhere.

The role of developed countries in tropical deforestation was first identified and described by François Nectoux, latterly in collaboration with Nigel Dudley, in a series of research reports from Earth Resources Research for Friends of the Earth. We are grateful to François and to the staff at Friends of the Earth who coordinated the work, in particular Charles Secrett, Jonathan Porritt, Koy Thompson, Andrew Lees and Simon Counsell. FoE's research work has been continued by Simon Counsell, Tony Juniper and Tim Rice, and has provided valuable information for this book. Other people who have contributed support, help and informed criticism over the years include Peter Bunyard, Mark Campanale, Julio César Centeno, Marcus Colchester, Mark Collins, Chris Cox, Pooran Desai, Barbara Dinham, Mary Edwards, Malcolm Fergusson, Don Gilmour, Herbert Giradet, Edward Goldsmith, Robert Goodland, Nicholas Hildyard, Simon Hodgkinson, Philip Hurst, Liz Hoskin, Julia Allen Jones, Scott Jones, Robert Lamb, Larry Lohman, Carlos Ibero, Hubert Kwisthout, Gerald Leach, Karin Lindahl, George Marshall, Gerry Matthews, Roger Moody, Norman Myers, Roger Olsson, Tim Peck, Duncan Poore, Kit Prins, Mike Read, Andy Rowell, Jeff Sayer, Tim Synnott, Tessa Tennant, Andrew Tickle, Bernadette Vallely, Lawrence Woodward and Rick Worrell. Other people should remain unidentified but have our gratitude. Particular thanks are due to Sue Stolton for help throughout the research and production of the book.

At WWF, we are lucky to be able to draw on the willing expertise of many people around the world, both in providing information and checking facts and opinions. We are grateful to people in the WWF Forest Advisory Group, and in other parts of the organisation, including Martin Abraham, 'Wale Adeleke, Charles Arden-Clarke, Clare Barden, Russell Betts, Monica Borner,

Debra Callister, Barry Coates, Tom de Meulenaer, Dominick DellaSala, Jean-Pierre d'Huart, Victoria Dompka, Chris Elliott, Cherry Farrow, Steve Gartlan, Arlin Hackman, Pierre Hauselmann, Merylyn Hedger, Ian Higgins, Stig Hvoslef, Thierry Jaccaud, Odette Jonkers, Lassi Karivalo, Harri Karjalainen, Mike Kiernan, Chris Laidlaw, Anders Lindhe, Paolo Lombardi, Tony Long, Isabelle Louis, Alison Lucas, Ian MacDonald, Eishi Maezawa, Adam Markham, Martin Mathers, Jill McIntosh, Günter Mertz, Anne-Marie Mikkelsen, Margaret Moore, Gonzalo Oviedo, Matthew Perl, Michel Pimbert, Steven Price, Yuriko Prod'hom, Mike Rae, Ulf Rasmusson, Per Rosenberg, Ugis Rotbergs, Jean-Luc Roux, Sarah Russell, Meri Saarnilahti, Sissi Samec, Gordon Shepherd, Shekhar Singh, Justin Stead, Magnus Sylvén, Richard Tapper, Georgia Valaoras, Nancy Vallejo, Olivier van Bogaert, Arnold van Kreveld, Pablo Xandri and Salahudin Yaacob. 'Wale Adeleke and Dominick DellaSala took the trouble to review the entire manuscript.

At Earthscan, Jonathan Sinclair Wilson, Jo O'Driscoll and Andrew Young kept polite and patient through numerous delays; we hope that the end result is worth the trouble. In addition, Earthscan's editor, Audrey Twine, did an excellent job in picking up factual and stylistic glitches in the text.

Despite our debt of gratitude, none of the above are responsible for any errors and opinions, which remain our own.

Preface

Early in 1994, the minutes of the UK Timber Trade Federation (TTF) referred at some length to the 'threat' posed by a book called *Bad Harvest*, to be published by Earthscan in association with the World Wide Fund for Nature (WWF). As the said book was, at that stage, little more than a three page memorandum and an unsigned contract, the fears seemed a trifle premature. Unfortunately, they are all too typical of the level of debate that continues to dog attempts at examining forest policy and the timber trade. If we were simply interested in writing an attack on the logging industry, then such attention might be flattering. In fact, we hope that this book will be more than just another hatchet job.

The text is divided into two main parts. The first (Chapters 1–6) looks at the role that the timber trade has played in the loss and degradation of forests around the world. Some of the arguments put forward here certainly *are* critical, as assumed in the leaked memorandum from the TTF. In opposition to some recent claims, we argue that in many ways the timber trade is the *primary* threat to many of the world's surviving natural and semi-natural forests.

The report draws on research that we have carried out over the last 10 to 15 years, working both individually and together. It summarizes and updates a number of previous papers and reports, written mainly for Friends of the Earth International and WWF. The text draws on unpublished data, academic books and papers, interviews and much personal experience. It also owes a great debt to the work of many friends and colleagues, in WWF, other non-governmental organizations (NGOs) and academic institutions, who have themselves studied the role of the timber trade around the world. While not trying to be comprehensive – a hopeless task for one slim volume – we aim to give a concise overview of the impacts of the timber trade on the global forest estate. We concentrate particularly on environmental issues, but also touch on social and ethical concerns where appropriate.

The second part of the book (Chapters 7–10) looks beyond the critique, to how things could be improved in the future. It starts by examining current attempts to address environmental and social issues through changing national and international policy initiatives, forestry practice and a range of market-based solutions. Here the critique is extended and includes casting a sceptical eye at how governments and environmental NGOs have addressed the issues. It then goes on to present a strategy relating to the timber trade and sustainable forest management, based on work carried out within WWF, and looks to a possible future, where forests are more effectively and equitably managed.

These issues are live. There is currently more talk, and more meetings and papers relating to the rather woolly concept of 'sustainable forest management' than there have ever been before. We have been lucky – or unlucky – enough to be involved in many of these debates, and try to present a guide to the maze. The speed of development means that the second part of the book will inevitably become dated before very long.

WHY THE TIMBER TRADE?

Foresters and timber importers sometimes ruefully ask why they have been singled out for criticism while other industries, such as metal mining and petroleum, seem to get off more lightly. To some extent they have a point; there certainly are many other industries that might be improved by some trenchant environmental campaigning. While oil drillers and miners increasingly face criticism for their role in degrading tropical forests, the lion's share of blame has been aimed at the timber industry. However, the timber trade* is unique in the extent and depth of its relationship with forests, as this book will illustrate. The threats to natural and semi-natural forests are clearly identified as amongst the most serious environmental problems of the late twentieth century, and as a result the trade has continued to come under the spotlight.

However, in anticipation of one of the timber industry's common complaints about environmentalists, we should stress that this book is *not arguing against the use of timber*. Timber from forests that are well managed, from a broad social and environmental perspective, remains one of the most environmentally-friendly products available. We are not suggesting that timber should be replaced by other materials, such as aluminium or plastic, which carry environmental costs of their own. It is a tragedy – and an avoidable tragedy – that at the moment most of the wood products on the market are not produced in a socially and environmentally sustainable manner.

Interest in these issues first emerged during a period of rapid growth in environmental awareness during the 1970s. At the beginning of the 1980s, some European environment groups began to take cautious steps towards a campaign seeking to highlight the problems caused by the rapid loss of

---

* We use 'timber trade' to refer to any business relating to timber, from forestry operations to high street retailers selling timber and paper products, and to both within country and international trade.

tropical rainforests. At that time, representatives of the European timber trade asserted that virtually no tropical timber was traded in Europe and that forest loss was due primarily to what were described as domestic problems such as 'overpopulation', agriculture and collection of fuelwood. Early research for Friends of the Earth put paid to these claims, by tracking tropical timber from natural forests to buyers in Europe, and by identifying companies involved in the trade. It showed that about a third of the world's internationally traded tropical timber** came to Europe, and that the trade was playing an important role in the degradation of several tropical forests.

Since then, similar studies around the world have identified the role of the trade in the degradation of many tropical, temperate and boreal forests. Over the course of 15 years, investigation of the timber trade's environmental impacts has been dominated by NGOs in both the North and the South. Today, almost everyone is familiar with stark images of forests under threat, ranging from the dense rainforests of the tropical belt to the remnant boreal forests of northern Europe and North America. Photographs of chainsaw-wielding loggers and clear-felling have become powerful apocalyptic images of the late twentieth century.

In the midst of the forests' retreat, a growing number of people in the rich countries are experiencing natural forests for themselves, through adventure holidays and independent travel. An emotional response in favour of forests is now increasingly assured from the public. However, there is still considerable confusion about why forests are under threat. While initial NGO activity focused simply on raising public awareness of the issues, along with industrial and political lobbying, present effort is increasingly directed towards searching for positive solutions to the urgent problems facing tropical forests. This includes work through both existing intergovernmental channels, such as the International Tropical Timber Organization (ITTO) and various United Nations bodies, and through involvement in independent initiatives such as the certification of well-produced timber, under the auspices of organizations like the Forest Stewardship Council.

NGOs stand at a turning point in their relationship with the timber trade. Damage to forests as a result of timber operations is now worse than it has ever been before on a global level. The anger shown by many protestors, both on blockades in the forest and during actions in the consumer countries, is testament to the depth of feeling about the fate of the world's forests. There is a conviction, which we share, that much of the timber trade is playing a pivotal and generally destructive role in the process of forest degradation and loss. At the same time, opportunities for better forest management and for timber certification mean that in some places old enemies are making peace

---

** Measured in roundwood equivalent (RWE), ie the amount of timber extracted from the forest to make particular products such as sawnwood. RWEs can be calculated for sawnwood, pulp, wood chips, etc and provide a more accurate measurement than simply volume of traded timber.

and working together towards common solutions to forestry problems. While NGOs and the timber trade in countries such as Canada and Papua New Guinea generally remain in conflict, and the battle lines have scarcely even been drawn up in the former Soviet Union, great strides forward have already been made in Switzerland, Finland, New Zealand and elsewhere. Discussions with representatives of retailers, foresters and state forestry bodies encourage us in the belief that, when the will is present, major improvements in policy are not as difficult to achieve as it sometimes seems on a day to day level.

We hope that this book will draw a reasonably accurate picture of the current turmoil, as well as presenting some facts and figures informing the debate. Our background is within the environmental movement, and our loyalties are clear. We remain frustrated, angry and sometimes appalled by the reckless selfishness shown by many people within the timber trade. At the same time, we acknowledge the real progress made by others who are forward-looking and responsible; they remain a minority but their numbers are growing.

We have worked, researched or travelled in most of the world's forest regions during the period that we discuss, and have been actively involved in campaigns against, and debate with, the timber trade. We have thus experienced at first hand many of the sights, arguments, losses, gains, and frustrations that currently bedevil efforts to improve management of the world's forest estate. We have attempted to distil this into a coherent picture and to point the way towards some options for improving forest management in the future.

The book has been written by Nigel Dudley, with active collaboration from Jean-Paul Jeanrenaud and Francis Sullivan.

Nigel Dudley, Machynlleth, Powys, Wales
Jean-Paul Jeanrenaud, Gland, Switzerland
Francis Sullivan, Godalming, Surrey, England
June 1995

1
A Vision for Forests

Let us begin on a positive note, by imagining a future where humans live in harmony with nature. What will this mean for the wood and wood products industry? Forests will be managed not just for timber, but also for the wealth of other goods and services that they can provide. They will be maintained as reservoirs for biodiversity. This does not imply that there will be no timber trade, but that the trade will be managed in a very different way than it has been until today.

While the debate about the future of the forests continues, and can still be highly polarized, some consensus is emerging about what constitutes good forest management. Different people's voices are heard, conflicts are brought into the open, and from this, a strong dynamic is developing to find positive solutions. It is hoped that we are moving away from a period of defining a problem and developing rhetoric, to a stage when genuine partnerships emerge to tackle problems and implement solutions: a movement from talk to action. There is now a general recognition on the side of industry that the environment is a serious issue, for both commercial and, increasingly for some companies, ethical reasons. Although this ideal still seems some way off, confrontation should become a thing of the past. The history of the human race suggests that emotion and aggression lead to conflict rather than to long-term solutions. The late twentieth century clearly illustrates the need for the human race to apply its collective reasoning powers to resolving conflicts rather than simply being swayed by prejudice and emotion.

Generally, the way forward should be through positive interactions, consensus building and attempts to find common ground and common solutions to problems of forest conservation and management. Elements in any such vision include the following (not an exhaustive list):

❑ **Responsibility.** The timber trade clearly has a responsibility – to shareholders, individuals, consumers and future generations – to manage forests for the long term and for the whole range of goods and services which they can provide. Responsibility rests on governments, corporations and, perhaps most importantly, individuals.
❑ **Partnerships.** One important way forward will be through the development of cooperation between many different interest groups.
❑ **Specific solutions for particular environmental and cultural**

contexts. At all levels, generalized solutions have been found wanting. In the future local people will have greater control over, and derive greater benefits from, management of local resources.

❑ **Building trust.** An important element will be the gradual development of trust between all the interest groups, many of which are currently in conflict. This will require adaptability and compromise from all participants.

❑ **Equity.** There will be equity between rich and poor countries, and between North and South, in terms of prices paid, conditions of trade, environmental controls, treatment of workers and standards of management.

❑ **Costs.** The true price, incorporating environmental and social costs, will be paid for wood and wood products. There will be a gradual shift away from a high volume, low value trade to a high value and generally lower volume trade.

❑ **Consumption.** As part of this development, there is a clear need to reduce wasteful consumption. Increased consumption, in terms of resource use, does not necessarily lead to improved social welfare beyond a certain point. There are limits to growth in resource use on a finite planet.

These are fine words. However, the vision remains a long way from the reality and the timber industry is threatening the future of some of the world's most fragile ecosystems. In the future, companies must ensure that the products they use and sell come from well managed forests. Any industry that ignores the sustainability of its raw materials supplies will become ever more vulnerable. Companies also have an obligation to conduct business in ways that minimize their impact on both environment and society. Governments, shareholders and customers increasingly demand openness and accountability. The next generation will judge industry against its present performance.

The authors believe that the wood and wood products industry has to face up to the challenge of responsible management of the world's forests, which constitute its resource base, and to work in partnership with environmental and other non-governmental organizations (NGOs) to ensure that products come from well managed forests – such a partnership is one important way of developing solutions to environmental and commercial problems. In the future, if problems are tackled together, there will be a greater chance of identifying solutions which will jeopardize neither industry nor the environment. This book explores some of these issues in greater detail.

2
A Global Forest Crisis

At the present time, the world's forests face two potentially devastating threats: a loss of total area of forest in large parts of the tropical and subtropical world, and a rapid decline in the quality of forest in much of the temperate and boreal regions.

Although far from precise, these categories help to identify and distinguish the historical changes sweeping through the global forest estate. They also show some of the key differences between the tropical, mainly developing countries, where total forest area is still often rapidly declining, and the rich nations of the temperate regions, where forest area is often increasing as forests recover from past over-felling but where management is intensifying and a 'forest' now often bears little relationship to the original ecosystem. These differences are not absolute. In some respects, forests in many tropical countries are coming to resemble those in temperate and boreal areas. Nonetheless, the two broad distinctions of *quantity* and *quality* remain a useful measure.

The situation at present may accurately be described as a crisis, and one in which time is a critical factor. More environmentally and socially equitable approaches to forest management are being developed in many parts of the world. If we were to be reasonably optimistic, we might argue that growing public opinion will eventually ensure that these methods are introduced, and continually improved upon, in the future. The current question is: how much high quality forest will survive the next few years, to ensure that ecological processes can continue, to maintain biodiversity and to allow maintenance and re-establishment of a balanced global forest ecosystem? Events in the next two decades will be critical in determining the richness and stability of many of the world's forests in perpetuity.

This book is about one particular agent for change: the international timber trade. However, to set it in context, this introductory chapter outlines key issues concerned with the importance, status of, and the threats to, forests around the world.

THE IMPORTANCE OF FORESTS

Forests cover much of the Earth's land surface. They perform a range of critical environmental and climatic functions and are home to the majority of the world's plant and animal species. For humans, forests have historical, religious, philosophical and aesthetic significance, as well as providing valuable resources and living space. Loss or degradation of forests therefore has a wide range of detrimental effects.

ENVIRONMENTAL SIGNIFICANCE

Forests form the natural climax vegetation over most of the planet's land surface, except in areas of natural desert, mountain tops, prairie, wetlands and the polar regions. They therefore have a key role in maintaining natural ecosystems. They act as natural regulators for a number of important ecosystem functions, including: maintaining and enhancing soils; reducing extremes of climate; controlling rainfall and hydrological systems; and acting as a reservoir for biodiversity.

Forests are important for maintaining *soil*, both by reducing rain and wind erosion and through building up the humus layer and improving soil structure by decomposition of dead leaves and wood. Soil erosion often increases dramatically when forest cover is removed, particularly in hill districts or where extremes of climate occur. Measurements in West Africa found that soil loss from cultivated fields was 6300 times greater than from tropical forests[1]. In Panama, which has suffered serious deforestation, some 90 per cent of the country is now affected by soil erosion[2].

Forest loss also increases *climatic extremes*. Deforestation can, paradoxically, cause both extra flooding and greater drought. Soil erosion has direct effects on hydrological cycles and the stability of watersheds, adding to the siltation of lakes and reservoirs and consequently heightening the chances of downstream flooding. There is evidence that water runoff increases with deforestation. Measurements in Indonesia found that runoff from logging roads exceeded 180 m^3/month, as opposed to 2 m^3/month from undisturbed forest[3]. About 60 million ha of India are now vulnerable to flooding, more than twice as much as 30 years ago, after a period of rapid deforestation[4]. Studies in India suggest that changes in transpiration following forest loss can result in a greater *intensity* of tropical rainfall, enhancing runoff and erosion, even if the total *amount* of rainfall remains unchanged. Forest loss can also make rainfall more erratic, thus lengthening the dry periods. During droughts there will be less water release from forests, causing desertification of humid and sub-humid zones[5].

The result of forest loss and climatic disruption in countries with a dry climate and poor soils is often *desertification*. Once a desert forms or expands it is difficult to restore the land to a natural or productive habitat again, because of changes in weather patterns, problems of establishing seed and the social and environmental problems that accompany desert formation, including loss of farmland. Desertification currently affects over 100 countries worldwide and is almost always connected with loss of tree cover. Asia and Africa each account for 37 per cent of desertified land, with other affected countries in Mediterranean Europe, North and South America and Australia. Over 30 million km^2 of land are currently suffering from desertification around the world[6].

Forests are also the world's most important reservoirs of *biological diversity*. The planet is currently experiencing a loss of biodiversity – in terms of species extinction, loss of genetic diversity and destruction of ecosystems – at a rate greater than at any other time in its history. According to E O Wilson, current rates of extinction are 1–10,000 times greater than background levels for the last half a billion years. It has been estimated that up to 50,000 species a year are becoming extinct[7] and further damage is occurring through loss of varieties and disruption of natural ecosystems. If current trends continue, 25 per cent of the world's species could become extinct, or reduced to tiny remnant populations, by the middle of the next century[8]. These figures remain controversial, but there is little doubt that biodiversity is declining fast. Most species are found in tropical moist forests, with other high 'islands' of biodiversity in temperate rainforests, so deforestation and forest degradation are significant in loss of biodiversity.

Tropical rainforests cover only about 6 per cent of the world's land surface, but are thought to contain between 10 and 50 million species, well over half the estimated global total[9]. Diversity is further concentrated *within* tropical moist forests to some extent. For example, it has been estimated that within ten principal biodiversity 'hotspots' – including Madagascar, western Ecuador, the Atlantic rainforest of Brazil, and Peninsular Malaysia – 350,000 animal species could become extinct within the next decade[10].

Areas of additional richness also exist within temperate forests, particularly rainforests. Biodiversity in some plant and animal groups, such as soil microflora and fauna, can approach that found in tropical forests[11]. Genetic diversity *within* species is also thought to be particularly important in temperate forests, making some local populations of great ecological importance. Pacific Northwest forests of North America have the highest conifer densities and greatest biomass in the world, and also the longest-lived species.

SOCIAL FUNCTIONS

It is ironic that, at present, most 'official' estimates of forest value rely on calculations of how much timber a particular area of trees can provide over time, the so-called *sustained yield*. Although these are the figures that gain headlines in the financial press, for most people the value of a forest is far more complex, and often in partial or complete opposition to its role as a supplier of timber and pulp.

At the most direct level, many people make their homes or living in forests. Indigenous people, often with little or no previous contact with the outside word, are sometimes wholly dependent on an intact forest ecosystem, and encroachment into this habitat can permanently alter or destroy their way of life. Many indigenous groups are currently threatened by changes to the forest ecosystem, including: the Penan in Sarawak, Malaysia; the Batak in Palawan, the Philippines; tribal groups in northern and northeastern India; the people of the Chittagong Hill Tracts in Bangladesh; the Waorani and other tribes of the Amazon; Inuit (Eskimo) people of Alaska and Siberia; Native American groups in Canada including the Nisga Nation, the Sekani, the Lubicon Lake Indian Nation and the Kyuquot People of Vancouver Island; the Sami people of Lapland, covering Norway, Finland and Sweden; and some aboriginal people of Australia.

Even where there are no aboriginal people, intact forests provide homes, resources, employment and recreational benefits. About 2 billion people, approximately a third of the world's population, rely on fuelwood as their primary source of energy, mainly collected from forest areas. Living forests provide employment, not only for loggers but also for the many people who utilize the resources in an intact forest, such as rubber tappers, collectors of Brazil nuts and other nuts, herbalists, people collecting rattan and bamboo, harvesters of berries and fungi, hunters and trappers, and those who manage a sustainable timber harvest through coppicing, selective logging, managed charcoal burning and small-scale timber production. For example, studies in the Korup Forest in Cameroon found that the value of the forest as a standing resource, for tourism, food supply, etc outweighed its potential value as timber. Pacific Northwest forests supply millions of dollars of non-timber products every year in the form of mushrooms, floral displays, fish and recreation. Research in northern Finland has shown that in some areas of the boreal forest, the value of berries and mushrooms is equal to 20 per cent of that provided by timber[12].

Forests are, however, of far more than simply utilitarian function. They also provide irreplaceable resources with respect to recreation, aesthetic pleasure, historical importance and cultural and spiritual associations. Many forests have local significances which cannot easily be recognized by people from outside. Concern for the future of high quality forests reaches well beyond environmental or conservation concerns.

STATUS OF GLOBAL FORESTS

Natural forests are in retreat throughout the world. In some areas total deforestation is taking place, along with the attendant social and environmental problems that this brings. In others, natural forests are being replaced by more or less inten-

sively managed secondary forests or plantations. Even where forests survive, the quality of many forests is declining because of a range of human influences including pollution, global warming and increased incidence of fire.

Although verifiable and comparable statistics about forests are notoriously hard to obtain, there are estimated to be about 3 billion ha of closed canopy forest in the world, along with a further 700 million ha of open canopy forest, 1.7 billion ha of other wooded land and about 29 million ha of plantations. This makes a global total of almost 5 billion ha of forested land[13]. About 850 million ha of moist tropical forest remain. Accuracy in measurement is not helped by the fact that the two principal United Nations agencies charged with assessing forest – the Food and Agriculture Organization (FAO) in the tropics and the UN Economic Commission for Europe (UNECE) in most temperate regions – have historically disagreed about their definitions of what constitutes a forest. Total forest area is divided roughly equally between tropical and temperate regions.

LOSS OF FOREST AREA IN THE TROPICS

Humans have long had an ambivalent relationship with forests. Trees play a major role in the spiritual and artistic heritage of almost all cultures, yet many societies have squandered their forest resources for timber, fuelwood or land development. Deforestation has been linked with serious social and economic problems for people as far apart as the Middle East, Greece, Iceland and the Easter Islands[14]. Industrialization has further accelerated the process of forest loss. Forests today cover only four fifths of the area that they did at the beginning of the eighteenth century[15], although this figure masks massive regional differences, with regrowth and reforestation with plantations in some places and a continuation of rapid deforestation in others.

Globally, tropical forests are undergoing the most rapid rate of change and over half the natural forests in the tropics have already disappeared during this century. An area of the Amazon rainforest equal to the size of western Europe has been completely deforested or seriously degraded during the last 40 years[16]. Annually, some 7.3 million ha are deforested in the moist tropical zone, 4.9 million ha in the tropical rainforest zone and about 2 million ha each in the dry and hill tropical zones[17]. Rate of loss continues to accelerate in many areas.

In some countries, destruction has been virtually complete. In the Philippines, over 80 per cent of coastal mangroves were destroyed between 1920 and 1980[18] and only a fraction of terrestrial forest remains intact. The latest estimates from the Côte d'Ivoire suggest that loss of closed forest is running at 6.5 per cent per year, despite the majority having already been destroyed[19]. Australia's rainforest, mainly found in northern Queensland, has largely been destroyed by logging and development, and considerable areas remain under threat from urban sprawl and holiday developments. Other countries, where substantial areas of rainforest still exist, are often undergoing rapid and sometimes accelerating forest loss. Examples include Indonesia, parts of Malaysia, Brazil, Cameroon and Zaire.

LOSS OF FOREST QUALITY IN THE TEMPERATE AND BOREAL REGIONS

Temperate and boreal areas are also undergoing rapid loss of natural forest. These effects are more difficult to identify because they are often followed by natural or enhanced regeneration. Although the area under trees remains the same, or may even increase, the natural forest is lost along with most of the biological and some of the social functions described above. Most temperate forest statistics do not distinguish between natural forest and secondary or plantation woodland, making it hard to track the changes.

Ownership of temperate and boreal forests is highly concentrated: the Russian Federation, Canada and the USA control over 70 per cent of the total, and the Russian Federation alone controls 41 per cent[20]. However, from an ecological perspective, some of the smaller temperate forest areas are critical resources of biodiversity, including for example those in parts of Europe, New Zealand, Australia and the tiny remnant temperate forests of South Africa.

In Europe, forest cover is estimated to be about 160 million ha, slightly less than half the original area. In western Europe it is estimated that the proportion of old-growth and semi-natural forest remaining is only 0.8 per cent of the original forest[21]. This is mostly in reserves or regions judged uneconomic for logging, although fragments of old-growth forest continue to be destroyed even here. Eastern Europe has more old-growth forest, although virtually none that has not been felled at some time in the past. In the USA, only 1–2 per cent of the original native forest remains, although proportions vary regionally so that Washington and Oregon states have 13 per cent old growth. The fate of these forests is the subject of enormous controversy. British Columbia in Canada still has almost 40 per cent native forests but these are rapidly disappearing, earning the province the title 'Brazil of the North' from some conservation groups. New Zealand retains 20 per cent of native forest; although some of this is regrowth from previously logged-over forests the majority is probably still primary or old growth. Estimates of Australia's remaining native forest cover vary from 5 to 20 per cent[22]. In some temperate areas of the South, there is still a net loss of forest cover. Chile still loses 50,000 ha/year and Iran 20,000 ha/year, although data remain rudimentary[23] for many areas.

QUALITY AND QUANTITY

While distinguishing between loss of forest quality in the North and forest quantity in the South is a useful shorthand for understanding the nature of the problem, it should not be taken too precisely. Increasingly, forest managers in the South are following northern forest management policies, and replacing natural forests with plantations, often of exotic species. Massive conversion to plantations is occurring, for example, in Malaysia and Indonesia. Loss of quality is thus becoming an ever more important issue in the tropics as well.

Loss of forest quality is also more complex than a simple measure of loss of area, and there are many concepts of what a 'good quality' forest should resemble[24]. Important elements identified by WWF International include: authenticity; forest health; environmental benefits; and other social and economic aspects.

These are discussed in greater detail in Appendix 1.

LOSS OF FOREST QUALITY

From both social and ecological perspectives, the replacement of natural or semi-natural forests with plantations or intensively managed forests results in a number of important losses. Management is arguably the most important single cause of loss of forest quality on a global basis (this issue is discussed in some detail in Chapter 5). However, in many areas the decline in quality is due to other factors, including pollution, global climate change and increased incidence of fire (summarized briefly below).

The impact of *atmospheric pollution* on tree health remains controversial. In the last decade, large areas of forest in Europe, North America and parts of Asia have shown symptoms of a novel form of decline. It is now generally assumed that this is due to a cocktail of different effects, including management practices, climatic variables and disease, but that in most areas air pollution also plays a key role, sometimes a dominant one, in the damage. In general, older trees are most badly affected, along with individual trees that are isolated or exposed at the edges of stands. Decline is often triggered by climatic factors including drought. Scientists have identified several features which, taken together, indicate loss of tree, and therefore forest, health. These vary between species but often include discoloration and cracking of leaves, premature leaf and needle fall, erratic branching of twigs and loss of crown density. Most commonly, damage includes general decline in vitality and loss of health; in some cases it can lead to death, either directly or by weakening trees and thus exposing them to pest and disease attack, or environmental stresses such as drought and cold.

Tree decline has been most closely studied in Europe. According to measurements by national assessors working to guidelines laid down by the UNECE and the Commission of European Communities (CEC), in 1992

some 23.5 per cent of trees were damaged in the continent as a whole, ie had
defoliation exceeding 25 per cent. Some 22.2 per cent of broadleaved trees
and 24.3 per cent of conifers in Europe were damaged, and the worst affected
tree species was *Quercus suber* with 32.7 per cent damage. The report concluded
that some countries 'regarded air pollution as the essential factor causing
forest damage in their countries. The majority of the remaining countries considered air pollution as a factor leading to the weakening of forest ecosystems
because of impaired nutrient uptake, increased soil acidification and reduced
base elements'[25]. Decline in Europe and the former USSR affected approximately 6 billion m^3 of timber. Problems are also increasingly being seen in
areas of Asia and the Pacific[26].

Air pollution is now also being linked with the possibility of more fundamental changes in climate through *global warming*[27]. A range of greenhouse
gases, and particularly emissions of carbon dioxide from burning fossil fuels
and plant material, are suspected of causing atmospheric changes that will
result in a gradual rise in average temperature. Current observed climatic
changes are similar to those predicted by computer models of global warming.
Although it will be some decades before we can be sure if these changes are
only temporary or mark a permanent shift in global climate, if they continue
forests in many areas will be affected[28].

Climate change could result in an overall shift in conditions suitable for
many temperate and boreal tree species, generally towards higher latitudes
and/or higher elevations, and new ecological communities could be created.
Under natural conditions, most tree species will be unable to disperse fast
enough to keep pace with the rate of climate change. Some predictions
suggest that up to 40 per cent of boreal forests may disappear[29]. Tropical
forests will also be affected. Increased hurricane frequency and intensity could
be a severe threat for susceptible forests. Marginal ecosystems, such as tropical
montane cloud forests, heath forests and tropical forests found at the edge of
their natural ranges (eg Xishuangbanna in southern China, or the forests of
Queensland in Australia), may be particularly sensitive to climatic changes[30].

Although *fire* is a natural component of many forest ecosystems, an artificial increase in forest fires can be ecologically damaging, and carries a heavy
financial and social cost. Worldwide, forest fires are becoming more frequent.
In 1988, almost 4 million ha were affected in North America[31]. In Europe, 90
per cent of forest fires occur in Spain, Portugal, France, Greece and Italy[32]. In
China, about 1 million ha of commercial forestry are destroyed every year by
fire[33]. Problems have recently been highlighted in Australia, where large fires
devastated forests and houses around Sydney in 1993. An enormous forest fire
on the island of Borneo burnt for several months during 1983: combined
effects of fire and drought burnt 25,500 km^2 of primary and secondary forest
and a further 7500 km^2 of shifting cultivation and settlements. Kutai National
Park was virtually destroyed by the fire, and in some dipterocarp forest areas
left unburnt by the fire 70 per cent of the bigger trees died of drought[34].

Although considered remarkable at the time, a pattern of major fires every few years has developed on the island, with the worst fires yet occurring in Kalimantan in 1994. Naturally occurring fires in West African savannah penetrate further into tropical moist forest once it has been partially logged[35]. In 1994, a vast fire on the Galapagos islands off Ecuador put several unique species at risk, including the giant tortoise (*Geochelone nigra*).

Some of the key impacts on the forests of the world are summarized in Table 2.1.

CAUSES OF FOREST DECLINE

People destroy or degrade forests because, for them, the benefits seem to outweigh the costs. Underlying causes include such issues as poverty, unequal land ownership, women's status, education and population. Immediate causes are often concerned with a search for land and resources, including both commercial timber and fuelwood. The impact of the timber trade is generally greater than has been claimed in the past. The North plays a key role in many of the factors leading to forest decline.

Forests are almost always destroyed or degraded for a reason. While we can criticize some of the reasons for the destruction or degradation of forests, and bemoan the effects, it is important to understand the web of different factors that can lead to forest loss. These factors can be divided into underlying causes and secondary or immediate causes. Currently, most studies of forest loss concentrate on secondary causes, such as the peasant with a machete, and ignore the underlying political, industrial and socio-economic causes that push the peasant into the forest in the first place.

Underlying causes
Underlying causes are those which help create the conditions in which forests are destroyed. Social conditions in both the North and the South are critically important. These include education, the position and status of women in society, levels of corruption and the degree to which government and industry are democratic and open to public inspection. The rich countries play a key role in setting the scene for forest decline through rapidly increasing consumption which in turn fuels a market for raw materials. It can be argued that the most significant 'population issue' relates to people in the rich countries, where every child born will use resources equivalent to those of 20 or 30 children in the poorer nations[36].

Other issues include pressure on land as a result of population growth and, often more important, inequality in land distribution in many countries. Inequalities of land ownership have gained most attention in Latin America, but affect land use in most areas of the world. For example, the second biggest landowner in the Amazon, Madeira Nacional SA, controls 4.1 million ha, an

Table 2.1 Selected crisis points in the world's forests

Continent or region	*Main locations of forest loss and its causes*
Europe	Forests of the **Mediterranean region** threatened by fire, development, tourism and some forestry Lowland conifer and broadleaf forests of **Scandinavia** now highly endangered from industrial forestry Upland forests of the **UK** at risk from sheep and deer grazing, tourism and land use change Wetland forests of **Latvia, Lithuania** and **Estonia**, logged for sale to foreign companies Old growth forests in the **Czech Republic, Slovakia, Poland, Bulgaria, Romania, Albania** and **former Yugoslavia** rapidly being felled for the timber trade
Africa	Tropical rainforests of **West Africa** rapidly depleted by logging and through clearance for agriculture, with the most threatened countries including **Nigeria, Cameroon, Côte d'Ivoire** and **Ghana** Subtropical dry forests in **East Africa** including **Eritrea, Ethiopia, Sudan, Somalia, Kenya** and **Tanzania** rapidly depleted by clearance for agriculture, fuelwood collection and overgrazing, and by war refugees Relic temperate forests in **South Africa** threatened with encroachment and climate change Forests in the **Atlas Mountains** of **Morocco, Algeria** and **Tunisia** under threat from grazing and logging
CIS	**Siberian forests** are planned for exploitation by transnational logging companies, **Karelian forests** are felled for export to Scandinavia and forest development planned for **Belarus, Uzbekistan**, etc
North America	**Pacific Northwest coast** forest is being logged out in the USA and Canada Increased logging for pulp in boreal regions of **Canada** **Alaska** rainforest logged by US and Japanese timber companies Key forest fragments, in **central Canada, Texas** and southern **USA** states, at risk of further losses Subtropical forests in **Florida** threatened by development and resulting changes in the water table
Latin America	Forests of **central America** are probably being cleared faster than almost anywhere else in the world, with those of **Costa Rica, El Salvador, Guatemala, Panama** and **Nicaragua** at particular risk Losses to the **Amazon** equal to the size of Europe from ranching and logging, and settlement continues

Table 2.1 continued

Continent or region	Main locations of forest loss and its causes
	Temperate beech forests of **Chile** and **Argentina** logged and replaced with pine plantations
Oceania	Eucalyptus forests of mainland **Australia,** especially in the south west, and in **Tasmania,** are being destroyed and replaced by plantations, and rainforest in **Queensland** threatened by development
	Tropical forests in **Papua New Guinea** and the **Solomon Islands** logged, including illegal operations
Asia	Himalayan region of **Nepal, India**, and **Bhutan** logged and degraded for fuelwood and fodder
	Lowland forest of **India** and **Nepal** rapidly, often illegally, logged
	Forests of **China** have been badly depleted and losses continue, especially within **Tibet**
	Rainforests and mangroves in the **Philippines** reduced to fragments but still being illegally degraded
	Forests in **Malaysia** and in **Indonesia** rapidly being cleared by farmers, loggers and for pulp plantations
	Thailand's forests have already been reduced to fragments over much of the country and a logging ban has increased pressure on neighbouring countries such as **Myanmar (Burma), Cambodia** and **Laos**

Source: Equilibrium (1995)

area approximately the size of Denmark. Another 1.6 million ha, originally controlled by an expatriate US citizen, and now in the hands of a Brazilian company, were bought specifically for forest production. In Scotland, the richest 1 per cent of the population own 52 per cent of privately-owned land and a further 22 per cent of such land is controlled by between 2 and 5 per cent of the population[37].

Underlying many of these factors is the world trade system itself, with its relentless and constantly shifting search for raw materials. Also important is the role of government and industry in promotion of international debt, which has forced many poor countries to liquidate their natural resources as assets to service their national debt repayments.

Immediate causes

Following on from the underlying causes of forest degradation are a range of 'immediate' causes; that is, causes which can be linked directly to effects within the forest. These include pressure from human settlement, the impact of industry, including the direct and indirect impacts of the timber trade, effects from other primary industries operating in forests, particularly mining, and the secondary effects of industrial pollution, especially of the atmosphere. The relative importance of these is hotly debated. A number of causes, such as air pollution, are outside the scope of this book. However, others need further examination, if only to distinguish them from the impacts of the timber trade.

Human settlement, and particularly *agricultural development*, is the principal immediate threat to many tropical forests. Over half the deforestation in the tropics is due to clearance for farming. International agencies, such as FAO, have in the past used such statistics to argue that other factors, such as the timber trade, are relatively insignificant. However, many settlers follow logging trails, originally come to the forest as loggers, or are forced into forests by demands on good land from other activities. This makes the underlying causes considerably more complex than they first appear.

Farming pressure takes a number of roles. In many areas there has been an increased rate of *shifting cultivation*, due to the exclusion of peasants from productive land, population growth and migration as a result of political persecution. The forest can often no longer recover between periods of farming. Migration of rice farmers into the northern uplands of Thailand has shortened the previously stable slash and burn rotation system from 8 to 10 years down to 2 to 4 years in places[38]. There has also been a rapid spread of *subsistence farming* due to permanent settlement, sometimes through forced or encouraged transmigration, or as a result of illegal encroachment. The Indonesian government's transmigration programme is the largest resettlement project in the world, moving people from the densely populated and fertile islands of Java and Bali to the outer islands of Kalimantan, Sulawesi and Irian Jaya. So far, about 4 million people have been moved, with assistance from the World Bank, Asian Development Bank and UK Overseas Development Administration. The project threatens 3.3 million ha of pristine rainforest, including tribal lands[39]. Irian Jaya (formerly West Papua) was illegally invaded by Indonesia and Indonesia's claims to the island are not recognized by the United Nations. Similar general policies have been directed towards the colonization of areas of the Brazilian Amazon. In addition, agricultural clearance plays an important role in the loss of temperate forest in some areas of, for instance, northern Asia and the Mediterranean basin. Past deforestation in many temperate regions has been driven by agricultural expansion, in some cases stretching back to prehistoric periods.

Development of large-scale farming, often ranching or plantations, has been particularly common in the forests of Central and South America. It is estimated that up to 1980 some 72 per cent of land clearance in the Amazon was due to expansion of cattle ranching[40]. The lifetime of many of the resulting cattle ranches is only 5 to 10 years[41]. Similar deforestation for ranching has taken place in Central America, including especially Costa Rica, which has also led to the further marginalization of peasant producers[42]. Unlike shifting or subsistence farming, large-scale agriculture usually produces products for sale in domestic markets or, more frequently, for export. Throughout Southeast Asia mangrove forests have been destroyed to create shrimp farms, although loss of mangroves damages indigenous fisheries by reducing the success of fish breeding. By taking up large areas of land, it sometimes also adds to pressure on remaining forests.

Another significant cause of forest degradation, and also a cause of deforestation around cities, is the use of *fuelwood and charcoal*. The importance of fuelwood collection as a cause of deforestation has sometimes been exaggerated, and it is often secondary in importance to agricultural clearance[43]. Nonetheless, searching for fuelwood is a time consuming and frustrating job for many people, principally women, in developing countries such as Nepal and India. When fuelwood is in short supply, in some countries dung is dried and used as fuel, thus further reducing soil fertility.

Compared with the impacts of agriculturalists, the effects of industry on the forests are often played down in international statistics and overviews[44]. It is true that a relatively small amount of *tropical* timber directly enters the international trade, and the amount is dwindling. Nonetheless, there are a number of ways in which industry, and particularly the timber trade, contributes to forest loss.

IMPACTS OF THE TIMBER INDUSTRY

To estimate the overall effects of the timber industry, it is important to distinguish both primary and secondary impacts of logging and forest management.

Firstly, and most importantly, industrial development is often the beginning of forest disruption. Loggers drive roads into previously impenetrable forests and at best selectively log trees, thus opening up forests for further exploitation by squatters, miners and would-be settlers. The timber trade brings towns, families, infrastructure and hangers-on, a proportion of whom stay behind to try their hand at unplanned development once the good timber has been logged out. The presence of logging roads has helped open up previously inaccessible rainforest areas in, for example, East Kalimantan in Indonesia, Thailand, the Brazilian Amazon, Ghana and northern Scandinavia. Railway developments have opened up untouched forests in Gabon and elsewhere. Sometimes loggers work together with other interests groups, such as oil drillers, on road systems, as in parts of Ecuador[45].

Secondly, statistics for the effects of the impact of the timber trade are, in

many cases, probably underestimates. Many national estimations are inaccu-
rate, both because they ignore many illegal incursions (which sometimes
exceed official logging, as in the Philippines) and because many official figures
are years out of date. Initial selection logging does not show up well in satellite
images; identification of later destruction through farming or secondary
logging will miss the connection with the timber trade. Loggers also penetrate
into some of the least disturbed forests in the world, threatening ecosystems
that would otherwise probably remain fairly intact. For example, logging is the
primary cause of forest loss in such important forest areas as Kalimantan, the
Central African Republic, Zaire, parts of the Amazon, the Pacific Northwest
of the USA, Alaska, Canada and northern Siberia.

Lastly, it should be noted that forestry, including both clearcut logging and
intensive selection management, is clearly identified as the *primary* cause of
loss of quality in most of the world's temperate and boreal forest ecosystems.
The timber trade's claim to be a negligible cause of forest loss is not supported
by the available facts. These issues are revisited in Chapter 10.

CONCLUSIONS

The loss and degradation of forests is one of the most critical environmental problems facing the world. Causes of loss are complex, and many start
well away from the forest itself, in the offices of governments and powerful
corporations. Far from being a negligible cause, we will argue in Chapters 3
and 4 that logging for the timber trade is the *primary* cause of forest degradation and loss in many of the remaining natural forests. Where the trade is
dwindling, it is not so much because of changes in policy, but because the
forest has been logged out as in, for example, the Philippines, Thailand and
the Côte d'Ivoire.

3
The Timber Trade: a Changing Global Structure

Changes in the structure of the international timber trade, and in forestry and timber technology, have major implications for forests. Issues include: a tendency to globalization of markets; a move towards fewer, larger companies; vertical integration within companies; and the increasingly important role of companies based in the South. These changes are in turn affected by wider political and economic developments. Changes in forest management and timber use also affect approaches to forests and plantations. Manufacturers are using a wider range of species, ages and qualities of trees, and demand is shifting from timber to pulp. As a result, the timber trade is able to utilize areas of natural forest that would previously have been uneconomic, and management of secondary forests is intensifying.

CONSUMPTION OF TIMBER THROUGHOUT THE WORLD

The market for timber has grown enormously in the last few years; for example between 1966 and 1988 world softwood production rose by 28 per cent and hardwood production by 54 per cent[1]. Global roundwood production in 1990 was 3.43 billion m^3, up from 2.93 billion m^3 in 1980. Industrial roundwood production was 1.66 billion m^3, an increase of 15 per cent during the 1980s. On a worldwide scale, over 50 per cent of timber removals are still used for fuelwood, mainly in the South. The North (mainly North America, Europe and Japan) currently consumes 76 per cent of total traded timber[2] and is also the major source of industrial timber, producing 1.26 billion m^3 in 1990, or 75 per cent of the total[3] (see Table 3.1).

Although the North's production approximately equals its consumption, this masks a growing internationalization in the timber trade. Many of the Newly Industrializing Countries of Asia, for example, now have extensive trade links (and ownership of land, concessions and companies) in the North. European and North American companies continue to buy timber from the tropics, and there is also an extensive and growing South-South trade[4].

Europe and Asia are net importers of timber, while Africa, North, Central and Latin America, Oceania and the former USSR are all net exporters. Regional totals obscure major differences between countries and, for example, Asia's position as a net importer is due to large imports into Japan, China, South Korea and Taiwan, whereas there are still major exports from, for example, Malaysia and Indonesia. During the 1980s, exports rose slightly in North America, Oceania and the Russian Federation, and by over six times in Latin America, but fell slightly in Africa. Europe and North America continue to dominate trade in timber and timber products[5].

Table 3.1 Production and trade in timber

Region	Percentage exports*
Africa	2
North and Central America	36
South America	4
Asia	15
Europe	33
Oceania	4
former USSR	7
World	101

* includes industrial roundwood, sawnwood, wood-based panels, wood pulp, paper and board

Source: *FAO Yearbook: Forest Products 1991* (1993) FAO, Rome

Global increase in timber use is not primarily due to population growth. The largest growth in consumption has taken place in the North, where population is relatively stable. Currently, over three quarters of the world timber consumption is in the industrialized countries[6], whereas the population of Europe and North America only makes up some 15 per cent of the world total. Reasons for the increase include instead such factors as: a shortening of product cycles (the 'throwaway society'); an enormous growth in use of pulp products for packaging and paper use; and changes in technology that have allowed wood fibre to be used for a wider range of purposes than in the past.

THE CHANGING FACE OF THE TIMBER TRADE

Over the past 20 years, profound changes in both the structure of, and attitudes within, the international timber trade have left critics of the industry with the impression that they are aiming at a constantly moving target. An analysis of trends that seems reasonable one year is shaky the next, and entirely out-dated in five years. Changes have affected the size of companies, the scope of their operations, attitudes towards forest resources and the technical options open to the trade. Implications for those working in the forestry sector are profound. These internal changes have their counterparts in wider social, economic and political reorganization throughout the world, including rapid economic growth among some of the so-called 'Third World' countries and the collapse of the Soviet bloc.

CONCENTRATION OF ECONOMIC POWER

Following a worldwide industrial trend, the timber trade has become increasingly concentrated in the hands of fewer, predominantly private, interests. Two factors are important:

❏ a tendency for large timber companies to grow even larger through swallowing up some of their smaller competitors;
❏ a withdrawal of state involvement in forestry in many areas, with a consequent shifting of resources to private hands.

The timber trade in some countries, such as the USA, Sweden and Japan, has long been dominated by a few transnational companies (TNCs). However, elsewhere trade was, until recently, run mainly by relatively small family firms, utilizing individually-owned forests. During the 1980s this long-established pattern began to break down. A series of mergers and acquisitions, often between companies based in different countries, enlarged the size and influence of the major corporations. Timber companies have often been assimilated into larger industrial concerns.

A few examples give the flavour of what is a worldwide phenomenon. The Finnish company Enso-Gutzeit Oy bought the French timber importer Becob in 1991[7] and tried to merge with another Finnish company, Kymmene, in 1992[8]. New Zealand's Fletcher Challenge has bought ownership or stakes in at least 12 companies since 1980, including: Crown Zellerbach Corporation of Canada in 1983; 87 per cent in British Columbia Forest Products Ltd (now Fletcher Challenge Canada Ltd) in 1987; 50 per cent in Australian Newsprint Mills in 1988; UK Paper plc in 1989; and forestry interests in Chile[9]. Another large New Zealand company, New Zealand Forest Products, was formed during the 1980s by acquisitions and mergers including Elders Resources and a 14.9 per cent stake in the Australian North Broken Hill mining company[10]. The Ireland-based chipboard manufacturer Finsa Forest Products is now owned by a Spanish group after originally being set up by a West German company and later taken over by an Irish company[11]. The UK company Reedpack was taken over by the third largest European company, the Swedish Svenska Cellulosa in 1989.

The process of merger and growth has been particularly marked in the pulp and paper industry. A string of acquisitions at the beginning of the 1990s significantly changed the balance of power in the market. Western companies began to explore the markets opened up by the collapse of communism and at the same time several Southern companies expanded rapidly, so that an Indonesian paper company, Indah Kiat Pulp and Paper Corporation PT, was ranked fifth largest in the world in the early 1990s, although its relative position has since declined[12].

This process has resulted in further concentration of market power in the hands of a few top companies. In Sweden, for example, three companies (Stora, SCA and MoDo) control 75 per cent of the paper market[13]. Despite the high profile given to Japanese timber companies in the media, it is the USA that has increased its dominance over the world paper industry, with six out of ten of the world's top companies by sales, following the most recent acquisitions; the others are based in Japan (2), the Netherlands and Sweden[14], (see Table 3.2).

Table 3.2 Ranking paper companies by turnover following acquisitions in the 1980s

The world's top pulp and paper companies are located in only a few countries in North America, Europe and Southeast Asia. Many are also successful timber companies. All the top paper companies have operations in many different countries, and their interests often span both tropical and temperate forests.

Ranking	Name	Country of origin
1	International Paper	USA
2	Nippon Paper	Japan
3	Georgia Pacific	USA
4	KNP BT	Netherlands
5	Scott Paper	USA
6	Stone Container	USA
7	James River Corporation	USA
8	New Oji Paper	Japan
9	Stora	Sweden
10	Mead	USA

Source: Pulp and Paper International (1994) 'Top 150 Listing' *Pulp and Paper International*, Sept 1994, Belgium

Increasingly, control of several companies in different areas of the world is seen as important, even if it is attempted only on a fairly experimental level. In Finland, Kymmene justified buying a minority share in the French sawnwood importer and distributor Becob in 1991: 'we believe that the better we can control our markets and get closer to the end user, the better we can develop our products'[15]. Kymmene also has a joint holding in a Uruguayan eucalyptus plantation, in partnership with Royal Dutch Shell[16]. The trade has thus changed fundamentally. Family firms and estates continue to be important in some countries, such as Indonesia, where in Kalimantan most forestry is controlled by about 250 companies owned almost entirely by ethnic Chinese Indonesians. However, the power of such companies to set the agenda of global forest policy is decreasing.

Privatization in central and eastern Europe

The other major change in ownership in some countries has occurred through complete or partial privatization of forest resources. This is taking place on a large scale in some of the Economies in Transition in central and eastern Europe, with results that it is still often too soon to predict with any certainty.

Privatization programmes have been initiated, and in some cases virtually completed, in most of the Economies in Transition in central and eastern Europe. In Latvia, it is planned that at least 50 per cent of forests will be handed back to previous owners or their families. Changes in the Baltic states led to a 727 per cent increase in volumes of timber produced there between 1993 and 1994[17]. Many of the formerly state-owned forests in the Czech Republic have been handed back to private ownership, with 144,000 new owners having an average of 2 ha. In Hungary at least 700,000 ha are marked for privatization and in Romania 400,000 ha are being distributed among a million peasant families, generally leading to rapid felling and sale of timber in a country where, unlike most temperate forest regions, total area of trees continues to decline[18].

Privatization has also taken place in a number of the Western economies, and in the South, as part of the emergence of 'free market development' that characterized the 1980s. In the UK, for example, the state Forestry Commission has been ordered by the government to sell considerable areas of its holdings to private investors. However, perhaps the most spectacular privatization to date took place in New Zealand in the late 1980s and early 1990s. Paradoxically, the sell-off came as a result of the avowedly socialist Labour Party's commitment to cut costs and streamline the sector.

New Zealand's giant sell-off

In 1985, the incoming Labour government of New Zealand planned a massive privatization programme, starting with its state forest industry, then estimated to be costing around NZ $1.5 million a week. In one of the largest land sales in history, senior executive Andy Kirkland dispensed with the majority of over half a million hectares of state forest in just a couple of years. Major beneficiaries were two New Zealand forestry companies, Fletcher Challenge and Carter Holt Harvey, which by 1987 together controlled 41.3 per cent of productive forest in the country. More controversially, major purchases were also made by foreign TNCs, including ITT Rayonier NZ with a parent company based in the USA, Juken Nissho of Japan, Earnslaw One of Malaysia and, through a joint venture, the China National Foreign Trade

Transportation Corporation. The US giant timber TNC Weyerhaeuser was initially interested, but eventually failed to invest[19].

Since then, foreign control of New Zealand's forest industry has continued apace. Carter Holt Harvey is now itself heavily influenced by foreign investors, with 11 per cent of its shares bought by the Templeton Fund in Hong Kong in 1991, and a further 16 per cent purchased by the US International Paper Group the following year. Over the same period, at least seven Japanese companies, including C Itoh, Sumitomo and Oji Paper, bought New Zealand forest companies. Several USA companies followed suit, and for example RII NZ Forests SI bought Tasman Forestry[20] in the early 1990s. An analysis of business opportunities in New Zealand states that:

> *the government of New Zealand has adopted a hands-off policy on the timber trade. It refrains from meddling in business decisions. It imposes no harsh restrictions on outsiders. . . What this means is that almost anyone can buy timberland in New Zealand. . . Foreigners can buy existing mills. . . or build new processing facilities*[21].

In Asia, a somewhat comparable trend is the public flotation of what have previously been private companies. Companies such as Barito Pacific and Odin have been floated, often on foreign stockmarkets, thus being transformed from national to transnational companies.

It is still too early to know how far many privatization programmes will go in practice. However, it is clear that in the future private capital will have an even greater control over the global timber market than it has done in the past.

VERTICAL INTEGRATION

Along with an increase in size is a spectacular broadening in the range of activities of the largest timber companies, in a process of *vertical integration*. Timber companies are being assimilated into larger industrial concerns which can control all stages of production, from growing timber in the forest, through processing and manufacture to marketing the end product in the high street. One result is that companies tend to see the forestry sector as just one part of their overall operations and to be much readier to move their operations around the world.

The example of Stora

For example, Stora, the giant Swedish TNC, claims to be the world's oldest existing company, tracing itself back to mining interests in the Middle Ages. Through a series of acquisitions, for example of Billerud, Papyrus, the Swedish Match and the German Feldmuhle Nobel, along with some strategic divestments, it has expanded considerably over the last decade. Group sales are around 70 billion SEK (£6 billion) and there are 70,000 employees. The group's activities are based on 1.6 million ha of forest in Sweden, along with 3800 GW of hydroelectric capacity. Stora's interests run from electricity to explosives, but centre on timber and its products, the company being involved in every stage of the production process from forest management (Stora Forest), through timber sales (Stora Timber and a range of marketing companies throughout Europe) to manufacture of a variety of papers (Stora Cell, Stora Papyrus, Stora Feldmuhle etc), packaging (Stora Billerud and Akerlund and Rausing), kitchen and bathroom equipment (Stora Kitchen), doors (Swedoor) and flooring (Tarkett, which has since been sold). In addition to European operations throughout Scandinavia and in Belgium, France, Germany, Ireland, Italy, Switzerland and the UK, Stora also has manufacturing units in Canada, Chile, India, the Philippines, Thailand and the USA. Total external sales in 1991 reached 72.71 billion kroner (£5.96 billion)[22].

Such a degree of control over all aspects of the trade has, in theory, allowed the larger corporations to mould all aspects of the timber trade into one coherent strategy. Forest management declines to one part in a much larger operation. This itself has a range of implications for the ways in which forests are managed.

THE PARTICULAR ROLE OF TRANSNATIONAL COMPANIES (TNCs)

The process of globalization of the timber trade means that the major participants in the industry are almost all TNCs. About 80 to 90 per cent of trade in forestry products is controlled by TNCs[23]. The increasing size, range and economic and political influence of these giant companies is of immense importance in determining policies towards forests around the world. Large areas of the global forest estate are therefore influenced by TNCs.

This has long been an acknowledged feature of forestry in developing countries, such as Chile where there are major holdings by European, Japanese, New Zealand and North American companies[24]. However, it is increasingly the case in northern areas as well. In British Columbia at the end of the 1980s, four major interconnected groups of companies controlled 93.2 per cent of the allocated public forest cut and the 20 largest companies controlled 74.1 per cent of the total cut[25]. Some 57 per cent of the control of public forests lay outside British Columbia, of which 36 per cent was outside Canada, including: 18 per cent in the USA; 2 per cent in Japan; 2 per cent in Finland; and 14 per cent in New Zealand[26]. Timber TNCs remain very active in many tropical countries as well. For example, in Côte d'Ivoire at least 17 foreign companies were operating in 1990, from France, Germany, Italy, the Netherlands and Denmark[27].

The largest TNCs now dwarf the economic power of many countries, particularly in the South. Even in the forestry sector, where there is usually a mixture of national, transnational and state companies, the economic power of the largest TNCs gives them a greater influence on markets and developments than that available to smaller companies or even state enterprises. This holds especially true for joint ventures between national or state bodies and foreign TNCs, currently seen for example in areas as far apart as the Russian Federation, Canada and West Africa. TNCs have helped to set the agenda for development policies within many Third World countries through their investment strategies, their influence on the pattern of aid programmes, and via related trusts and research organizations. They also wield an important influence on government and domestic industry through patronage, effectively buying favours through gifts, entertainment and sometimes also access to travel.

The environmental and social impact of the timber trade is closely tied up with its structure. This structure is fundamentally dependent on the existence of TNCs. In its final submission to UNCED, the UN Centre on Transnational Corporations (UNCTC) stated that 'the specific roles and responsibilities of transnational corporations ensue from their unique management aspects, the breadth of their corporate networks, the range of their technological resources and the international consequences of their decision-making'[28].

TNCs can also shift their operations around the world, to take advantage of new or cheaper resources, cheaper labour, fluctuations in the global currency exchange, tax breaks and less stringent health, safety and environmental standards. While it is unclear to what extent this occurs in practice, many TNCs admit that they do not necessarily adopt the same level of environmental controls in all their operations[29]. For example, Stone Container Corporation of the USA, which is trying to develop a major wood-chip mill in Costa Rica, has admitted that it engages in projects in Central and South America to avoid more stringent US environmental policy[30].

INTRA-FIRM TRADE, TRANSFER PRICING AND ILLEGAL OPERATIONS

A feature of TNC operations which has particular significance with respect to the timber trade is the opportunity for intra-firm trade within TNCs. Although intra-firm trade and transfer pricing can take place within a multi-plant national firm, it achieves particular significance in the context of TNCs because of their frequent variety of operations and the fact that deals are carried out in several different currencies. Controlling prices within an umbrella group of companies gives many opportunities for such activities as reducing the tax burden by declaring profits in a lower tax country; taking advantage of favourable exchange rates; reducing the proportion of profits going to subsidiaries in joint ventures; under-reporting profits where foreign investment is a sensitive political issue; and generally increasing flexibility of trade. Such light financial footwork also provides the opportunity for illegal trade operations. Although it is certainly not the case that the TNCs are the only companies involved in illegal forestry operations, their economic dominance means that a proportion have been linked with unlawful logging and extraction.

The case of Papua New Guinea

There have, to date, been few thorough investigations into the actions of foreign timber companies in tropical forests. One notable exception is the Commission of Inquiry into Aspects of the Timber Industry in Papua New Guinea, established in May 1987 by the Prime Minister of Papua New Guinea and chaired by Judge Thomas Barnett. The inquiry was established following persistent criticism of foreign-based timber companies operating in Papua New Guinea, including a report from the UNCTC, which hinted at transfer pricing and reported that 'Despite a flourishing timber trade, it was not until 1986 that any logging company declared a profit in PNG'[31]. The findings of the 'Barnett Report' provide compelling evidence that transnational timber companies have been acting not only in an environmentally damaging manner, but also illegally in parts of Papua New Guinea. Some extracts from the interim and final reports present a disturbing picture:

> *It would be fair to say, of some of the companies, that they are now roaming the countryside with the self-assurance of robber barons; bribing politicians and leaders, creating social disharmony and ignoring laws in order to gain access to, rip out and export the last remnants of the province's valuable timber. . . .*
>
> *These companies are fooling the landowners and making use of corrupt, gullible, and unthinking politicians. It downgrades Papua New Guinea's sovereign status that such rapacious foreign exploitation has been allowed to continue with such devastating effects to the social and physical environment, and with so few positive benefits. . . .*
>
> *It is doubly outrageous that these foreign companies. . . have then transferred offshore secret and illegal funds. . . at the expense of the landowners and the PNG government*[32].

The following quotation from the report refers specifically to Gaisho (NG) Pty Ltd (a wholly-owned subsidiary limited of Gaisho Ltd of Osaka, Japan):

> *Gaisho's other faults were dwarfed by the enormity of its illegal marketing strategies. Gaisho (NG) has a very complex marketing strategy and unravelling it has been like peeling an onion. The peeling process has disclosed true market prices as being 20–22 per cent above the price which the PNG producers received from Gaisho*[33].

Although the Commission of Inquiry was initiated by the government of Papua New Guinea, evidence of widespread abuse of power, including corruption among cabinet ministers, meant that Justice Barnett received steadily less support as the investigation continued. He was stabbed, nearly fatally, outside his Port Moresby home. Only two of the five interim reports have been printed, on a very limited run, and none is now publicly available in Papua New Guinea. Little of the evidence uncovered in the inquiry has been used to either halt individual companies or change practices. One of the people accused of serious corruption later became Deputy Prime Minister[34]. Since the Barnett inquiry, the dominant role in Papua New Guinea forestry has switched from TNCs based in Japan to others based in Malaysia and it is thought that one Malaysian company now controls 86 per cent of logging. But there are as yet only very limited signs that conditions are improving. A new Papua New Guinea forestry minister, who acted to prevent illegal logging, has recently been removed from the post. A new forest Policy and Forestry Act was passed by parliament, in response to the Barnett Inquiry and a World Bank Review. However, there is little evidence that anything has changed. Log exports were limited to 3.5 million m^3 in 1994, although this offers little hope of a reduction as it is 250 per cent higher than 1991 log exports and 140 per cent higher than the generally accepted maximum sustained harvest level. They also rely on loggers' own submitted tallies. Investigations by the Pacific Heritage Foundation found widespread evidence of continuing corruption, including by government ministers. Of one operation, where three oil palm projects were established by Singapore-based companies, including one of 500,000 ha, Max Henderson, director of the Pacific Heritage Foundation said (in a presentation at Oxford University in July 1994):

> *It is very difficult to visualize an equivalent. Perhaps if 50 French bulldozers were to appear totally without authority one morning in the New Forest, chain saws barked, and logs were loaded on jinkers and driven to the beach, you may have some idea of what is happening frequently in Papua New Guinea and other Pacific nations.*

The specific issue of illegal trading will be returned to in Chapter 4.

CHANGING OWNERSHIP PATTERNS

In addition to enlarging their market share, the internal structure of TNCs continues to change. One of the most important developments is an increase in the number of shareholders in many companies, as private companies are launched onto the stock-market and shareholder buying patterns change. Today, large timber corporations are likely to have more shareholders than in the past, both in terms of private individuals and through pension funds, where money is invested second-hand for a large number of policy holders. Ownership may well slip out of the hands of the home country. Indeed, companies are increasingly looking to the global market for investment. For example in 1993, P T Barito Pacific, a large and controversial Indonesian company that is involved in native-forest logging in Indonesia, was floated on the London exchange in a fanfare of publicity.

Multiple ownership has several effects on companies. In some cases an increase in anonymous investors, simply interested in returns on money invested, further reduces controls over TNC operations as executive staff are left with a sole mandate of maximizing profits. On the other hand, some shareholders, and particularly those with an interest in maintaining a good public image such as pension funds and other investment companies, can be relatively susceptible to outside criticism of a company and themselves exert pressure for change. To a large extent, the long-term implications of these social and environmental changes have still not been tested.

THE IMPORTANCE OF TIMBER TRADING

Tracking the amount of timber traded in national and international markets is frequently complicated by poor records, confusing statistics and in some cases by the amount of illegal trade. Total volumes of traded materials, as recorded by the FAO in its annual *Yearbook of Forest Products*, gives an indication of trade, but can be confusing in terms of total timber extracted from the forest. Comparing cubic metres of logs with cubic metres of veneer, for example, omits the considerable volume of timber lost or discarded during the production of veneer. Greater accuracy is possible by converting all timber to Roundwood Equivalents (RWEs), although this system remains very approximate and the proportion of timber wasted during processing varies enormously between countries and individual companies. Identification of timber species is even more approximate, and the majority of customs statistics, the main source of trade data, use only the most approximate means of classifying timber types. At one time, for example, several hundred species from at least three continents were being classified as mahogany, although only one genus, *Swietania*, is known as mahogany to botanists.

Nonetheless, a picture of the importance of trade can and is being developed. *International* trading actually makes up a relatively small proportion of the total timber in the marketplace. However, the global trade includes sources from some of the most highly endangered forest ecosystems, in Latin America, Asia, the Russian Federation and North America. *Within country* trading is increasingly dominated by the larger timber companies as well.

THE ROLE OF THE TIMBER TRADE IN THE NORTH

The developed countries of the North import nearly 80 per cent of all internationally traded timber products, and export 90 per cent of timber products except for non-coniferous sawlogs, veneer logs and sawnwood, veneer and plywood; ie, most trade is to and from the industrialized countries. Domination by the temperate countries is particularly striking in the case of coniferous logs, with the USA exporting more than half the world's total in 1989 and other major exporters being the former USSR, Canada, New Zealand and Germany. Canada dominates the coniferous sawnwood exports, while Australia and the USA dominate wood-chip exports[35]. Some examples of the structure and importance of the trade are given below, concentrating on Canada, Finland and Australia.

Canada

Among major forest nations, Canada is unique in that 94 per cent of its forests are publicly owned. The remaining 6 per cent belong to more than 425,000 private landowners. However, much of the forest is leased out to major companies, and the distinction between state and private industry is sometimes blurred. Total forest estate is 416 million ha, or 45 per cent of land area. About 46 million ha, or 11 per cent, are protected from harvesting, while somewhat over half the total estate is considered potentially to be commercial forest. Forest types are classified as softwoods (64 per cent), hardwoods (15 per cent) and mixed forests (21 per cent). In 1992, 168.3 million m^3 were extracted[36].

Over the past 20 years, Canada's share of total wood exports has ranged between 18 and 23 per cent of the world total. In 1991, wood product exports were worth Can \$20.6 billion, major components being wood pulp (24 per cent), newsprint (28 per cent) and softwood lumber (24 per cent). The majority of exports (66 per cent) went to the USA, with other major markets in the European Union (EU) (15 per cent) and Japan (9 per cent)[37].

Finland

Although Finland has some of the world's largest forest products companies, most forests remain in the hands of small private landowners. Forest covers 20.1 million ha, or 66 per cent of the land area, and consists of 62 per cent Scots pine (*Pinus sylvestris*), 27 per cent Norway spruce (*Picea abies*) and 8 per cent broadleaved trees, mainly birch (*Betula* spp). Ownership is divided between private owners (63.1 per cent), the state (28.1 per cent, mainly in the far north) and large companies (8.8 per cent). Some 300,000 people each own over 5 ha of forest, although there are also many smaller forest plots around summer houses, and about a million people, or a fifth of the population, own some forest. Average size of forest estates is 35 ha. Companies have been prevented from accumulating land because laws favour local people when forest is sold.

Forest products had 33.4 per cent of the export share in 1993, proportionately down from 90 per cent in the 1950s. Timber has been overtaken by the metal industry, but remains the biggest foreign income earner. Today 50 million m^3 of timber are cut every year, and there are 30 million m^3 of excess growth. Average size of clearcut in southern Finland is now 1.5 ha. Most conservation areas are in the north.

Industry infrastructure includes 43 pulp plants, 29 paper plants, 16 paperboard plants, numerous sawmills (of which 120 control 90 per cent of the market) and 38 plywood, particle board and wallboard plants. Paper production has increased; in 1970 some 4.3 million tonnes were produced as compared to 10 million tonnes by 1993. Exports are concentrated in Europe (78 per cent, of which 66.1 per cent is to its EU neighbours), Asia (9.2 per cent) and North America (6.3 per cent). The UK bought 25 per cent of exports 20 years ago, but the proportion has since declined to 16.7 per cent. Germany is another major market, taking 17.6 per cent of the total in recent years. Currently, some 12 per cent of the timber used is imported; 80 per cent of this is Russian birch but imports also come from Estonia, Poland, Germany, Scotland and Brazil[38].

Australia

Australia has about 43 million ha of forest land – covering 5 per cent of the total land area – which is about 62 per cent of the forest existing at the time of European settlement. In addition, there are 92 million ha of woodland. About 60 per cent of the forest is dominated by various species of *Eucalyptus.*

The forest estate is divided almost equally between four different classes of owner: private owners (11.3 million ha), state forests (11.5 million ha), conservation reserves (9.8 million ha) and other crown land (10.6 million ha). Within the state forest, about 7 million ha are available for logging, along with 6.8 million ha of other crown land and 7.6 million ha of private land, although not all of these reserves are economically worth harvesting. Native forests are still being logged and there is net forest loss. There are also 1 million ha of plantations, mainly of *Pinus radiata*, and 30,000 ha are established each year.

The wood products industry employs about 40,000 people, or 0.5 per cent of the workforce. Some 11.5 million m^3 of hardwoods and 6 million m^3 of softwoods are produced every year, with half going for veneer and sawlogs while the rest is used for pulp. Australia exports about a third of its annual output, mainly as hardwood chips to Japan, with trade in 1991–92 worth Aus $407.6 million[39].

THE GROWING ROLE OF THE SOUTH

At the present time, the timber trade is therefore still dominated by companies based in the North. However, over the past decade there has been a rapid growth of size and influence of key companies located in the South, and particularly in the booming economic areas of Southeast Asia and Latin America. Although still mainly minor exporters on an international stage, these companies have particular significance through their frequent role in exploitation of natural forests. Many act wholly as importers, but in some important exceptions they are directly involved in logging, both at home and abroad.

TIMBER TRADING IN SOUTHEAST ASIA

Southeast Asia is rapidly becoming one of the key centres for the international timber trade. Japan plays a critical role in both the tropical and the temperate timber trade, and has tended to dominate discussions about the Asian trade. However South Korea and Taiwan, both importing about US $1 billion worth of timber a year[40] are also significant on the world stage. Other Southeast Asian countries are also importing more wood products, both because of exhaustion of domestic timber sources and to meet a growing need for raw materials, particularly pulp. Thailand and the Philippines now have some controls on trade in their own timber, due to deforestation, and are importing timber from other Southeast Asian countries, particularly Indonesia and Malaysia. Imports of logs to the Philippines grew from zero in 1987 to 400,000 m^3 in 1990, with much coming from New Zealand; Thai imports have shown an even steeper rise, after a logging ban was introduced in 1989, and reached over 1 million m^3 a year by the 1990s[41]. Thailand is important because many of its imports come from Vietnam and, illegally, from Myanmar (formerly Burma), Cambodia and Laos. Hong Kong and Singapore, despite their small size, both import substantial quantities of timber, again from the two big exporters in the region, Malaysia and Indonesia. Both are also significant re-exporters through manufactured goods[42]. Two of the more significant importers, South Korea and Taiwan, are discussed in more detail below.

South Korea

South Korean industry is dominated by a few large trading groups, known as *chaebol*, most of which are involved in the timber trade. There has been a recent increase in timber imports and, for example, between 1987 and 1988 there was a 14.6 per cent rise in industrial wood production[43]. The USA was the largest source of raw logs in 1989, supplying 45 per cent to Malaysia's 40 per cent, and other major sources included New Zealand, Papua New Guinea and Chile. Malaysia and Indonesia supply most of Korea's sawnwood, and

Canada and the USA supply 85 per cent of pulp[44]. South Korea is one of the main centres of imports (along with China) of coniferous sawlogs, mainly from North America[45], and also imports timber and pulp from Latin America, principally from Chile.

Korean industry has become involved in direct logging operations. For example, a Korean company, Sungkyong, is running a large-scale forestry operation in rainforest in northwest Guyana in a joint venture with the Malaysian Sam Ling Timber. Another Korean company, Hyundai, has been involved in Siberian logging. The Borneo International Furniture Company has been active in Indonesia, although the latter's ban on raw log exports has now forced the company to diversify its supplies[46]. Other South Korean companies apparently operating in Indonesia include Leewan Industrial Korea, Hanni Industrial Co, Codeco Co Industries, Sungkyong Ltd, Taneco Lumber Ltd and You One Construction Co. South Korean companies are also involved directly in Malaysia, including the Sam Ho Timber Co Ltd in Sarawak. There are 108 pulp companies owning 120 mills. Most are small, although 26 larger mills exist[47].

Taiwan

Since 1946, when a right-wing rearguard army retreated from Mao Tse Tung and took over the island of Formosa, renaming it Taiwan, the state has remained in implacable opposition to communist China. Neither government recognizes the other and states have to choose recognition of one or the other. In recent years, a tendency to switch to recognition of China, for example by South Korea and Indonesia, has increased Taiwan's isolation. Even international trade figures are scarce, because United Nations bodies link China and Taiwan together as one. Nonetheless, Taiwan has maintained one of the most booming economies in Southeast Asia, including a regular trade with China, albeit carried out covertly through intermediary countries.

Within the forestry sector, some 90 per cent of the island's needs, about 6 million m^3, are met by imports, mainly from Southeast Asia and North America. Imports rose during the 1980s, declined as a result of recession, and now appear to be increasing again[48]. Malaysia is the single largest source of timber, supplying 80 per cent of log imports in 1990, of which 2.98 million m^3 out of a total 3.3 million m^3 came from Sarawak[49]. Taiwanese timber manufacturers are reported to be interested in investments in Malaysia, with the All Best Corporation investing M $10 million to expand forestry in Kuantan in the state of Pahang[50]. In 1990, the USA supplied 201,000 m^3, mainly from southern pine concessions obtained by Taiwanese interests in Kentucky, although the proportion of US timber is falling.

Other suppliers included Thailand (62,000 m^3) and Papua New Guinea (45,000 m^3)[51]. Plantations in the Philippines supply wood chips to Taiwan[52]. A Taiwanese firm interested in establishing itself in Canada was recently criticized for pressurizing the Alberta provincial government to guarantee that

timber supplies would not be impaired by future conservation legislation[53].
Rattan imports are also substantial, averaging 2500 to 3000 tonnes/year.
Currently, Vietnam and Papua New Guinea are the major suppliers, having
recently surpassed Indonesia. Fibreboard is increasingly imported, with major
suppliers including the USA, New Zealand, Australia and South Africa[54].

Role of the timber trade in the Economies in Transition

The future role of the former Soviet bloc countries of the Commonwealth of
Independent States (CIS), and of central and eastern Europe, is the wild card
in the pack with respect to futures speculation in international timber trading.
After years of steady, if unspectacular exports (considering the size of the
overall resource) there are now signs of an expansion in cutting in some areas,
while in other places the timber trade appears to be declining as a result of
political and economic instability. Several major changes can be seen:

❑   the demise of the Council for Mutual Economic Assistance (COMECON)
    trading partnership has damaged or destroyed traditional trading links in
    the area, thus radically altering patterns and directions of trade flows;
❑   privatization of land means that forest resources are becoming available to
    private individuals, and in some cases to both national and foreign
    companies;
❑   inefficient state forest bureaucracies are being replaced by private entre-
    preneurs often intent on capitalizing their assets as quickly as possible; and
❑   widespread social and political breakdown, particularly in the former
    USSR and former Yugoslavia, is resulting in additional opportunities for
    illegal felling and trading.

The political changes are still so new that trends have yet to show on
published statistics, which often lag at least two years behind reality.
Production in the Russian Federation until 1991 appears to have been fairly
constant, although this may mask a substantial increase in illegal trade.
Despite fears of a sudden rush to sell timber by new owners, the apparent
short-term effect in many European Economies in Transition has in fact been
a *decline* in timber exports[55]. Throughout the 1980s, many central and east
European countries underwent major year-by-year fluctuations in trade,
including particularly Poland and the former Yugoslavia, although Hungary
increased its cut of industrial roundwood and Romania its wood pulp production. Illegal cutting is reported to have risen sharply in Romania, although
much of this was for fuelwood during an energy crisis and did not enter the
trade[56].

GOVERNMENT AND INTERGOVERNMENTAL LINKS WITH THE TIMBER TRADE

As the influence of the major timber trading corporations grows, in both proportional and real economic terms, individual governments have attempted to exercise control over timber resources. In most cases, this has involved introduction of controls on forestry management, logging levels and the form in which exports take place. Many countries now have environmental and social elements included in their forest management policy, although the extent to which these are enacted in practice varies widely. Government controls also include such policies as the logging ban introduced in Thailand, a ban on logging certain species as practised in Ghana and a ban on exporting any timber except plantation-grown trees as practised in the Philippines. Other controls are probably aimed more at developing local industry, such as the ban on exporting raw logs from Indonesia, which has resulted in a rapid increase in domestic processing, although this has been matched by a collapse amongst small processors in countries such as Japan and Taiwan. Intergovernmental regulation is becoming to some extent more wide-ranging, eg the voluntary controls suggested by the International Tropical Timber Organization (ITTO), discussed in more detail in Chapter 7.

There is a lively debate about the efficacy of such controls. In a separate development, they are coming under attack through the fall-out from the latest set of negotiations of the General Agreement on Tariffs and Trade (GATT), the Uruguay Round and the establishment of the World Trade Organization. GATT negotiators have argued for some time that such bans are an impediment to free trade and, for example, the European Community (now European Union) challenged the Indonesian raw log export ban under GATT agreements in the early 1990s[57]. The free trade philosphy of GATT has been enthusiastically embraced by members of the timber trade, who fear the threat of official bans and boycotts on wood products from particular areas. In a long attack on environmental NGOs and a defence of free trade at the Third Global Conference on Paper and the Environment in 1995, Geoffrey Elliott, Vice President of Corporate Affairs for Noranda Forest Inc of Canada, concluded that:

> *The forest products industry . . . share(s) a common interest in maintaining open international markets for trade in pulp and paper products. To this end, pulp and paper producers should work together to preserve the integrity of the open global trading system as embodied in the rules based WTO/GATT.*
>
> *Proposals for changes in existing trade rules and related* initiatives to promote the use of trade measures as a mechanism to enforce environmental progress pose new threats to the integrity of the open global trading system *[our emphasis]*[57a].

TECHNICAL CHANGES WHICH HAVE AN IMPACT ON FORESTRY

Until relatively recently, many countries still relied predominantly on loggers with hand tools to fell trees, on animals to drag them out and on rivers and railways to transport timber. Over the last few years, a revolution both in forestry management and in the ways in which timber and wood products can be used has had wide-ranging implications for both the global forest estate and the people who make their living out of forestry.

Changes in the forest

Perhaps the two most important changes within the forest have been the construction of more roads and the development of mechanized cutting equipment. (Roads are discussed in detail in Chapter 4.) Mechanization, including initially the chainsaw, and heavy machinery for lifting and transporting logs, have revolutionized forestry. Although major logging is possible without such equipment – as demonstrated by the widespread clearance in forests in India during the eighteenth and nineteenth centuries and earlier deforestation of much of Europe and the Middle East – the chainsaw has provided a new impetus for logging, including illegal logging, in many parts of the world. It has speeded up the process and thus the economic returns, and has been instrumental in the opening up of many tropical forests.

Despite being identified by the public as a symbol of forest destruction, the hand-held chainsaw is already obsolete in many areas. Further innovations, both in the sophistication of forestry operations and in plant breeding, are allowing a totally new type of forestry to be practised. Mechanization has led to development of integrated harvesting machines which can carry out felling, sawing and transport of logs far more cost effectively than traditional forestry gangs. Other relatively new methods include whole-tree harvesting techniques, improved storage and drying processes, and improved methods of chipping, including residue chipping[58].

IMPACTS ON FORESTRY WORKERS

These changes have a radical effect on employment in forestry in most temperate countries. As yet, cheap labour means that mechanization has been less important in the tropics, although this situation will change. During a period when forestry operations have grown in many countries, numbers of workers have fallen dramatically over the last few decades. Sometimes, when a new piece of equipment is introduced, jobs can disappear almost overnight. For example, one modern harvesting machine can replace 8 to 10 chainsaw operators. In Sweden it is estimated that numbers of foresters have decreased by 90 per cent in the last 30 years, and chainsaw workers became superfluous in many forests at the end of the 1980s as a result of new harvesting equipment being introduced[59]. In Canada, the logging industry employed 1 per cent fewer people per year throughout the 1980s, culminating in a drop of 6.7 per cent in a single year, between 1988 and 1989, a year when revenues from the forestry sector rose by 39.6 per cent[60]. In the UK, the number of forestry workers fell from around 13,000 to 7000 between 1960 and 1980, despite a virtual doubling in area of managed forest[61].

This means, incidentally, that although the risk to employment is often used as an argument against protecting more native forests by the timber industry, in most places mechanization has put far more people out of work than conservation legislation. Employment trends were analysed in a Wilderness Society study looking at the US Pacific Northwest. Between 1980 and 1988, timber industry employment in the region fell by 14 per cent, while output of finished lumber increased by 19 per cent. If planned changes in the timber industry continued, an estimated 33,600 timber-related jobs, more than a third of current levels, would be lost through changes in mill efficiency alone. Proposals to increase the area under conservation management, by contrast, would affect less than a third as many people[62].

In 1993, Champion International Corporation abruptly announced that it was pulling out of a huge forestry operation in Montana, after years of cutting at a rate that would have been illegal in public forests. Dr Thomas Power, chairman of the economics department of the University of Montana in Missoula said: 'Champion came in here promising that they would be here forever, and then just overcut all the trees and left'. James Hill, leader of the Western Council of Industrial Workers, whose members faced a layoff of 1500 persons, added: 'I've been in the timber industry since 1951, and this is the biggest single blow I've seen – far worse than any cutbacks from environ-mental restrictions'[63].

Changes in tree breeding

Innovations are also taking place in tree breeding, particularly through advances in biotechnology. At present, the main technical development has been in the use of tissue culture methods to clone an almost unlimited number of trees from a single plant. Further progress through the use of genetic manipulation is expected in the medium term. There have been many questions raised about the ethics, safety and efficacy of genetic engineering. From the perspective of this book, the main impact of these new develop-ments will be concentrated in the field of employment. Further job cuts are likely and at the very least there will be a continued shift in *where* people are employed within the wider forest industry.

IMPROVEMENTS IN EFFICIENCY OF TIMBER USE

Technical modifications further down the chain of timber use also have direct effects on overall demand. Innovations in sawmilling, including com-

puterization of operations and introduction of such resource-saving innovations as saw dry rip (SDR) and edge, glue and rip (EGAR), have improved the
efficiency with which raw timber is processed. Changes in pulp processing
have had a dramatic effect on the amount of fibre feedstock required per unit
of wood pulp over the past two decades. From a forestry perspective, technical
innovations have changed the hardwood component in pulp; in the 1950s this
would usually have been no more than 10 per cent of total feedstock, but is
now often up to 40 per cent in Europe[64].

Changes in the end use of timber products

Perhaps even more important from the standpoint of forest management are
the changes in the end uses of timber. In general, the demands of manufacturers and consumers are moving away from high quality timber (especially
sawn logs) and towards cellulose pulp and wood chips used in a variety of
reconstituted building and manufacturing materials. For these products, technological developments mean that there is less restriction on the species, age
and quality of timber used as raw material.

Timber use has changed rapidly over the last few centuries, and each
change has had an impact on forest management. The needs of iron smelting
and shipbuilding, for example, had an enormous effect on the development of
forest management policies in Europe, including development of the *coppice
with standards* system. At the beginning of this century, fuelwood was still a
major component of timber use, responsible for 136 million m^3 of wood in
1913 as opposed to 54.5 million m^3 of sawnwood[65]. This use declined shortly
thereafter, with more timber being used as sawnwood for building, then later
as plywood and, as manufacturing techniques improved, in a range of other
particle boards, fibreboards and pulp products. These changes are being
mirrored in other countries around the world. Over 35 per cent of world
timber production now goes to pulp, and this is expected to rise to 50 per cent
by the year 2000[66]. Already over 50 per cent of European timber is used for
pulp[67].

Pulp looks set to dominate the world timber markets. Advances in polymer
resins have allowed development of stronger and more versatile reconstituted
wood products, which are replacing raw timber across almost the whole range
of uses. Reconstituted materials are used in building (using moisture-resistant
medium density fibreboards), furniture-making, packaging and DIY (home
build) applications. Long and extremely strong laminated softwood beams can
be made using multiple small pieces of wood bonded with high performance
polymer resins[68]. Composite materials, incorporating plastics, insulation and
decorative surfaces are also becoming more common.

Even the market for high quality timbers is being affected by technical
developments. Recent advances in Japan allow logs to be 'cooked' in a
microwave oven and compressed into a square shape as they cool, producing
a material which is stronger, denser and less liable to split or warp. Five litres

of water are squeezed out of a typical one metre long cedar log by the process, giving it the strength of the more expensive Japanese cypress[69].

This has striking implications for forestry practice. Instead of having to balance quality against quantity, modern timber merchants can aim for quantity alone, and rely on processing to make up the losses in strength, appearance and durability. From the perspective of forest management, this means that:

❑ many more tree species can be used, for example uses of aspen in boreal regions have recently increased;
❑ trees can be grown for maximum fibre production over time, leading to such proposals as five-year rotation, densely planted 'forest plantations';
❑ poor quality trees can be harvested that would otherwise have been left;
❑ whole tree harvesting becomes more attractive;
❑ and because of the capital-intensive nature of much of the equipment used, this adds a further impetus to centralization of power within the industry, as referred to earlier in this chapter.

CONCLUSIONS

The changes discussed above have had several important interconnected effects: increased pressure on native forests; increased intensification on secondary or managed forests; increased penetration of the forest market by TNCs; and increased globalization of trading in timber, pulp and cellulose products. The implications of these changes take up much of the rest of the book.

4
Logging in Natural and Semi-natural Forests

The global timber trade is in a state of transition between logging natural and semi-natural forests and extracting timber from managed forests and plantations. However, a significant proportion of the timber trade apparently intends to continue native-forest logging, for short-term economic gain. The damage that the industry continues to inflict on remaining natural forests is currently the most urgent forest conservation issue. This chapter examines the politics of the logging debate, discusses environmental impacts of logging, looks at the importance of the timber trade in the destruction of natural and semi-natural forests, and identifies where these pressures are most significant. Some case studies illustrate key issues.

The most urgent, and certainly the most emotive, environmental issue relating to the timber trade concerns the logging of natural or semi-natural forests. Despite increasing areas of managed forests and plantations, much of the world's timber continues to come from primary, natural or old-growth forests[*], ie from forests where human intervention has been minimal or has occurred in the distant past, and where natural ecological processes continue essentially unaltered. The implications of this logging pressure for conservation of biodiversity, for environmental stability, and even for the long-term sustainability of the global timber market, are profound.

THE POLITICAL DEBATE

The global forest industry is in a state of transition between logging natural forests and managing timber plantations. The industry likes to present this as a planned changeover; in fact the change is *ad hoc*, sporadic and as a result of past over-cutting rather than any planned process. The next two or three decades, at most, will see an end to the logging of natural forests in most

---

[*] Definitions of widely-used terms such as 'natural' and 'old-growth' remain vague, or vary in different parts of the world.

parts of the world. Whether this will come as a result of changing management policies, through conservation initiatives and stewardship of the global forest estate, or because almost all natural forests have been destroyed, is an issue that has occupied the attention of the conservation community for some years. As things stand at the moment, it looks as if the last will be true in many places: natural-forest logging will stop only when there are no more natural forests to log.

This situation has not come about by chance. Under conventional economic analysis, natural and semi-natural forests are consistently undervalued, so that it is far cheaper for a timber company to log a natural forest than to establish a plantation. It is also cheaper to establish a plantation on the site of a standing forest, thus getting in addition the value of timber from the natural forest, than to establish plantations on degraded lands. The full forces of late twentieth-century economic theory and practice are stacked against the survival of extensive areas of natural forest, anywhere in the world. This means, furthermore, that *any* natural forest is now fair game for the industry. Far from being a minor cause of forest loss, as is often claimed, the industry is the *primary* cause of natural-forest loss in the temperate and boreal regions and a *major* cause everywhere else. Public opinion is consistently on the side of greater protection, and is consistently being ignored. It is against this rather gloomy backdrop that conservation organizations, concerned governments, consumers and forward-looking members of the timber trade, are trying to develop forest management strategies that allow space for natural forests.

To a large extent the survival of natural forests has entered the endgame; timber traders are rushing to extract as much cheap timber as possible before conservation legislation, and sometimes explicitly with the aim of 'liberating soil' from natural forests for plantation establishment. Conservation organizations, indigenous peoples' groups, concerned individuals and a steadily growing band of direct action teams are struggling to stop the destruction of remaining natural forests. Winners and losers in this long war of attrition will be measured, fifty to a hundred years hence, by the proportion of natural forest that remains.

Here is the real front line in the forest debate. On one side are the conservation activists, blocking logging roads, hugging trees, lobbying governments and organizing petitions, often backed up by sympathetic media and significant sections of public opinion. On the other side are timber traders, logging companies and often the forest workers themselves, arguing that forest management is essential for jobs, economic security and resources, and that modern logging techniques allow a sustainable yield of timber and cause little long-term environmental damage. Although generalizations are difficult, these interests often have the backing of the financial press, industry bodies and often also the majority opinion in the national or regional government. Depending on the public relations skills of the industry lobby, they may or may not also be backed up by a fair proportion of the general public, who have been convinced that environmental concerns are exaggerated.

The feelings raised about the future of natural forests are among the most emotive in the whole environmental debate, reminiscent of the passionate campaigns for and against whaling. Anyone who has experienced the genuine anger on both sides of the argument, in countries such as the USA, Canada, Australia, Malaysia and Brazil, is likely to feel depressed about the opportunities for rational argument and balanced decisions.

In the Pacific Northwest of the USA, where the survival of the northern spotted owl (*Strix occidentalis caurina*) has become a *cause célèbre* in the old-growth logging debate, loggers and their supporters drive cars with bumper stickers reading 'I like spotted owl pie' and the head of public relations for a major industry group likened 'preservationists' to 'subversive elements. . . . the same people who opposed the Vietnam War and are in favour of abortion'[1]. Conservationists, on the other hand, have spiked trees to prevent chain saws cutting through them and sometimes seem to be prepared to use the full weight of the law to protect any and every forest. The two sides – and they are sides – usually meet only in situations of confrontation. There appears to be little common ground for even starting a debate.

Yet in some countries a measure of agreement has been reached between foresters and conservationists. In New Zealand, this has even culminated in the signing of a joint Forest Accord between most leading conservation groups and the principal industries concerned. In Switzerland, conservationists and the government increasingly agree about the direction of forest policy. Unfortunately, these examples are still the exception rather than the rule. Getting to the bottom of the native-forest logging debate is fundamental to determining the overall effect of the timber industry on the global environment.

THE IMPACT OF LOGGING ON SOCIETY AND THE ENVIRONMENT

Why is logging of such concern? The industry claims that it is simply the first stage in a sustainable forest production cycle. Forests have been felled for timber throughout history, and timber company spokespeople have argued that logging operations pose few long-term threats to the environment. Few people seriously suggest that we should stop using timber, so is there really a problem at all? Before looking at where, and how much, logging takes place, its ecological and social significance needs to be assessed.

Issues of concern relate to what type of logging is practised (see box opposite), and to when and where it takes place. Researchers trained in ecology, anthropology and social sciences have found abundant evidence that inappropriate or careless logging causes a range of harmful impacts. These are often most acute when logging takes place in natural or semi-natural forests. However, inappropriate logging practices can cause a series of harmful knock-on effects even in intensively managed forests and plantations. From an environmental perspective, impacts can be divided into a number of areas of concern: loss of

TYPES OF LOGGING

There are a number of different methods of harvesting trees, ranging from complete removal of all trees to selective removal of a few individuals or parts of trees. The method chosen has a profound effect on both the end product and the ecological value of the remaining forest.

Methods include:

Clearfelling

In this method, all the trees in an area are removed. The area clearfelled can vary from very small plots to large forests of many tens of hectares. Two main categories of clearfelling can be distinguished:

❑ whole tree harvesting, where the entire tree is removed, including all twigs, roots and foliage, usually for woodchips or fuel;
❑ timber removal, whereby trunks and major branches are taken and remaining twigs etc are either left as brash or burnt.

Selective felling

Here, only certain trees are removed. Selective felling embraces various different methods including:

❑ removal of a certain number of trees from a given area, retaining a relatively unbroken canopy;
❑ removal of one or more commercially important species (as commonly used in many parts of the tropics);
❑ removal of certain age classes of all trees (as used in some traditional European selective logging systems);
❑ removal of some parts of trees without killing them, thus allowing regrowth, through:
  – coppice, ie regular cutting at the base of tree to promote shooting and regrowth;
  – coppice with standards, ie a coppice wood with some full sized trees growing;
  – pollards, ie regular cutting near the top of the trunk to create bushy growth of new shoots;
  – selective removal of branches, for fuelwood use;
  – selective removal of other parts of the tree, such as nets, leaves (for fodder), bark (as in the case of cork oak) etc.

Source: Adapted from Dudley, N (1992) Forests in Trouble WWF International, Gland, Switzerland

natural habitats and biodiversity; direct and indirect effects on human communities; disruption of environmental stability and ecological processes;
disturbance of the climate hydrological cycles and watershed stability; and
invasion of pristine environments. Each of these issues is examined in more
detail below.

Loss of natural habitats and biodiversity

Logging destroys natural habitats, and this in turn results in the loss of biodiversity, sometimes leading to local or global extinction of species. Once trees
are removed, many associated life-forms can no longer survive.

Until recently, concern has focused particularly on the tropics, where the
huge number of locally endemic species means that disruption of even quite
small areas is potentially damaging. Tropical forests are thought to contain
over half the world's total species, somewhere between 10 and 50 million in
all, most of which have never been described[2]. The extent to which logging
has so far actually threatened species, or local varieties, in the tropics remains
the subject of debate, although there is almost unanimous agreement that
major losses will occur. Professor Edward O Wilson, of Harvard University,
assuming an annual reduction in area of tropical forest of 1 per cent,
estimated a consequent species loss of 11 to 16 species/day[3]. Some scientists
question the accuracy of such predictions, and believe them to be overesti-
mates, but few deny the reality of current losses of biodiversity. A series of
articles collected and published by the International Union for the
Conservation of Nature and Natural Resources (IUCN) in 1992, while being
more cautious, estimated that by 2040 some 17 to 35 per cent of tropical forest
species could be 'committed to extinction' as a result of all forms of tropical
forest loss[4], but did not distinguish the logging component from this figure.

Temperate and boreal forests also have important biodiversity values, particularly in temperate rainforests. Biodiversity in some plant and animal
groups, such as soil microflora and fauna, can approach that of tropical
systems[5].

One problem in trying to assess likely impacts of logging on biodiversity is
that our knowledge base remains incomplete, even in the most carefully
studied ecosystems. In more remote forests, the majority of species have never
been described by science. During the time of the writing of this book, for
example, two new mammal species have been discovered in the forests of Asia,
in areas under threat from commercial logging.

Furthermore, our understanding of the *nature* of biodiversity remains
tenuous. For example, the Canadian timber industry likes to point out that
biodiversity, measured in terms of numbers of particular plant species,
increases following a clearcut[6]. However, this is due to a sudden growth of
more widespread pioneer (and 'weed') species, and masks losses in overall biodiversity on a landscape level. The importance of biodiversity is thus less
about total *number* of species than about the presence of the *authentic*, often

rarer, species for the particular ecosystem, and of species representing all the stages in the natural cycle of the forest. Species likely to suffer most acutely from logging are often those associated with the oldest age classes of forests. Examples include large carnivores, which have large home ranges, low fecundity, and require undisturbed roadless areas.

Old growth in the Pacific Northwest USA

The crux of conservation arguments about logging old-growth forest in the US Pacific Northwest is that certain wildlife species are dependent primarily on old growth to survive. Brown (1985) lists 77 wildlife species that use old growth as their primary breeding ground, and 65 other species that use it as their primary feeding habitat[7]. The President's Forest Plan evaluated effects of ten forest management options on over one thousand species associated with old-growth in the Pacific Northwest.

Three bird species have received particular attention. The northern spotted owl is dependent on large tracts of old growth for nesting and foraging. Recent Forest Plans have been said to threaten an estimated 30 to 50 per cent of the birds through habitat destruction in the next 50 years[8]. The pileated woodpecker (*Dryocopus pileatus*) nests in large diameter (old-growth) snags and feeds largely on insects found in large quantities in dead logs. Again, 50 per cent or more of its habitat could disappear over the next few decades. The marbled murrelet (*Brachyramphus marmoratum*) is a small auk, spending the winters at sea and coming ashore to breed in the coastal old-growth forests. The first North American nest was not discovered until 1974, on a moss covered Douglas fir in northern California[9]. Biologists believe that this species, which is already rare, needs old-growth to breed successfully. The northern goshawk (*Accipter gentilis*) is also dependent on old forests.

Although the 'flagship' species like the spotted owl get all the attention, the survival of other species is also at risk. Often threats to smaller or more obscure species can exceed those to mammals and birds. Researchers in the Pacific Northwest have found that composition of fungi changes dramatically after clear-felling[10]. For example, the 'most noble' polypore, *Oxyporus nobilis-simus*, is probably the world's largest fungus. The biggest specimen known weighed 300 lbs (136 kg) and measured 1.4 m across. Only four or five specimens of this fungus have ever been found, and it is believed to be entirely confined to the ancient forests of the Pacific Northwest[11]. Arthropod species found in the canopy are also far more diverse in old growth than in younger forests[12].

The stable habitat created by temperate old-growth forests also provides benefits for species not directly dependent on the habitat itself[13]. For example, the decline of salmon species in the Pacific Northwest has been linked with logging and the subsequent decline in water quality due to sediment from roadbuilding and clearcutting. In November 1991 the sockeye salmon (*Oncorhynchus nerka*) was registered as an endangered species in the region. Paul

Brouha, executive director of the Association of Professional Fish Biologists, told three House subcommittees that forest practices under the plans were likely to contribute to 'the decline and extinction of native fishes over vast portions of their range'[14]. Public lands in the Columbia River basin contain 76 native salmon populations at high or moderate risk of extinction[15]. The American Fisheries Association estimates that 214 native salmon populations are at risk of extinction in the Pacific Northwest[16].

Indeed, as research continues, the news seems to get worse rather than better. Recent studies indicate that populations of deer and elk, which had previously been assumed to benefit from the open spaces created by clear-felling, are also likely to suffer declines as the old growth is destroyed, not least because they will no longer have access to cover from hunters or thermal cover in the winter. Particular fears have been expressed about the future of the Columbia white-tailed deer (*Odocoileus virginianus ieucurus*) populations in the Roseberg and Lower Columbia River areas, and about the Roosevelt elk (*Cervus elaphus roosevelti*)[17].

Similar threats can be seen in other temperate areas. Ingmar Ahlen, professor of ecology at the University of Umeå, Sweden, believes that 40 vertebrate species which either breed or feed in Swedish forests are now seriously endangered, along with 50 species of fungi, lichens and flowering plants[18]. For example, the Siberian tit (*Parus cinctus*) population has been reduced to 10 per cent of its original numbers following heavy logging in northern and boreal forests[19].

One particular cause for concern is *fragmentation* of forest habitats. Fragmentation occurs when felling takes place in small and discrete areas, as in much of North America, through road building, large scale timber harvest or because all but tiny reserved areas are felled. Species can become trapped in old-growth fragments. Even some bird species may find it difficult to disperse across clearcuts and for example they increase risk of predation of juvenile spotted owls, and many invertebrates and plants are confined to diminishing islands. The roads accompanying logging are a major cause of decline in large carnivores such as the grizzly bear, lynx and wolverine.

Clearcuts expose remaining trees to more wind damage and to harsher climates. Studies of two major blowdowns in the Mount Hood National Forest in the USA found 48 per cent and 81 per cent respectively took place beside clearcuts and roads[20]. It is estimated that microclimate can change significantly up to 400 m back from the forest edge in the Pacific Northwest. Greater fragmentation of forests and the creation of small islands is thus even more destructive to 'deep forest' habitat, which may not survive undisturbed inside the remaining fragments. Research on fragmentation in tropical forests, while in many cases less advanced, also predicts increased, although idiosyncratic, extinction and impeding of succession[21].

Recently, the industry in the US has used fears about 'forest health' as a justification for extra logging. 'Forest health emergency areas' are proposed for any area where fifty per cent of trees are dead or damaged as a result of disease or fire, or are susceptible to such damage. Ecologists point out that this could include almost any forest, and that this could be used as a way of felling areas that have, until now, been protected[22].

Direct and indirect effects on human communities

Many people live and work in forests, and logging almost always has impacts on their lives. Most of the world's largest intact or near-natural forests are still homes for indigenous groups and tribal people. Some tribes have inhabited and used forests continuously for millennia, and have well developed living systems including an intimate knowledge of the relationship between various plant and animal species, the properties of a large proportion of the plants that surround them and sustainable ways of collecting food, materials and herbal medicines.

Indigenous people are likely to suffer from any encroachment of their traditional lands. Logging operations are not infrequently their first contact with the outside world. Impacts range from direct persecution to introduction of diseases and destruction of habitat. Many tribes are currently, or have recently been, engaged in conflicts with the timber industry. Examples include the Dayaks and Penan in Sarawak, Malaysia; the Batak in Palawan, the Philippines; tribal groups in northern and north-eastern India; the Waorani and other tribes of the Amazon; and many Amerindians in Canada and the USA.

Campaign groups such as Survival International have focused attention onto events in the tropics. In Brazil, for example, 14 Tikuna Amerindians were murdered by loggers in 1988[23], and the assassins are still free because of the power of local logging companies[24]. Three Yukpa Amerindians were killed by loggers in Venezuela in 1994[25]. Amerindian people continue to be murdered by the military in Guatemala; for example, 13 people were shot during a peaceful protest against incursion into forest lands in December 1990[26].

However, these problems are by no means confined to the tropics. Important groups connected with temperate and boreal forests include Inuit (Eskimo) people of Alaska and Siberia and Amerindian groups in Canada including the Nisga Nation, the Sekani, the Lubicon Lake Indian Nation and the Kyuquot People of Vancouver Island. The last named have already been moved once to make room for timber operations and are now directly threatened by nearby clearcutting[27], while almost all those listed above have legal battles, and sometimes blockades, aimed against logging. The Sami people of Lapland, covering Norway, Finland and Sweden, have been in conflict with timber companies over old-growth logging in areas where they graze reindeer. In New Zealand, the Maoris have a continuing land-claim dispute, and some aboriginal people of Australia are protesting against logging on traditional sites.

Indigenous people do have fundamental and internationally agreed rights to continue to maintain their cultures and ways of life unimpeded by outside interference. These are recognized under international law and under the Rio

Principles drawn up at the Earth Summit (UNCED, 1992). Unfortunately, these rights are frequently ignored in practice. Gaining recognition for rights has become a legal minefield in a number of countries. Many indigenous groups have never made treaties with colonizers. Some, in the past, were verbally promised that their sovereign status would be recognized but cannot provide written proof of this. However, in other cases groups retain their original rights of occupancy and community title to forests, and logging operations are then a direct infringement of their rights. This has seldom stopped logging in practice, even in the rich countries of the North, although the situation is now gradually changing. Some tribal groups have also been associated with destructive logging activities.

Logging affects other people living in forest areas as well, through the environmental changes that take place, through impact on fisheries, reduction in the tourist trade, and simply through loss of landscape values and a historical link with the forest. These 'side issues' are often omitted from consideration in planning forestry operations and in the calculation of costs and benefits from logging.

Disruption of environmental stability

The second major issue with respect to logging is the range of direct and often economically quantifiable environmental effects that occur as a result. These are summarised in Table 4.1 and include soil erosion, landslips, changes in water quality and damage to hydrological systems and watersheds. Logging and replanting tends to produce uniform stands leading to loss of genetic variation. There are concerns about the depletion of nitrogen and other nutrients from the most intensive forestry systems.

Soil erosion increases dramatically when forest cover is removed, particularly on steep slopes and in hill districts. Measurements in West Africa found that soil loss from cultivated fields was 6300 times greater than from tropical forests[28]. In Panama, which has suffered serious deforestation, some 90 per cent of the country is affected by soil erosion[29]. In 1975, a severe storm affected the Mapleton District of the Siuslaw National Forest in Oregon, USA, and 245 landslides were documented. Of these, 9 per cent were natural events due to the storm, 14 per cent were related to road construction and 77 per cent occurred in clearcut areas[30]. A study of soil moisture levels for 20 years after clearcutting in the Douglas-fir (*Pseudotsuga menziesii*) forest of the Oregon Cascade Range found marked effects, varying from an initial increase in soil moisture levels to a long-term decrease, creating possible moisture stress for young trees[31]. In late 1993, Fletcher Challenge, the New Zealand forest TNC, faced three charges under provincial fisheries and forestry acts in Canada, which charged that harvesting and road building had triggered a landslide on Queen Charlotte Island in 1992. It was claimed that fish habitat had been destroyed[32].

Erosion has direct effects on hydrological systems and watersheds, increas-

Table 4.1 Impacts of logging

Impact	Notes
Clearfelling	Impacts on plants and animals through habitat loss
	Long-term loss of habitat for many smaller animals and lower plants, risks of local extinctions
	Risks of soil erosion and landslips during storms
	Problems of siltation and increased flooding in rivers, leading to reduction in fish numbers
	Local climatic impacts
Selective logging	Damage to remaining trees, leaving them open to pest and disease attack
	Opening up forest to migrants and illegal settlers
Fragmentation	Isolation of species in natural forest fragments and problems of reproduction, migration, etc
	Impacts within remaining forest from windblow and climatic influence
Road construction	Major impacts on soil erosion and degradation of watersheds during construction and throughout use
	Increasing fragmentation of forest ecosystem through creating barriers that some plants and animals are unable to cross
	Providing entrance into pristine forest areas for migrants, illegal settlers, etc
	Providing pathways for feral animals, pests and diseases, which further disrupt the natural ecosystem in remaining forest fragments

Source: Equilibrium (1995)

ing siltation of lakes and reservoirs and downstream flooding. About 60 million ha of India are now vulnerable to flooding; this is more than twice as much as 30 years ago, over a period when deforestation intensified[33]. Effects are also seen in temperate countries. Logging affects watersheds, creating torrent debris and risking sheet erosion[34]. However, one detailed study in the Pacific Northwest suggests that the impact appears far less severe than in the case of tropical monsoon watersheds and that the many different factors involved are poorly understood as yet[35].

The result of forest loss in countries with a dry climate and poor soils is the beginning of the desertification process. Once formed, it is difficult to restore to a natural habitat again, due to changes in weather patterns, problems of establishing seedlings, and the social and environmental problems that attend desert formation.

Climatic effects

There is good evidence that logging on a large scale can also disrupt local and perhaps also global climate patterns. At least half the rain that falls on the Amazon is recycled through transpiration, and the same water often falls several times on the forest as clouds pass over the continent. Cutting the forest can be expected to have severe impacts on weather patterns. When Malaysian rainforest was cut to create rubber plantations, the clouds stopped at the edge of the remaining jungle. Current problems of drought and flooding in, for instance, India, East Africa and Thailand have been linked to deforestation and the resulting climatic changes[36]. It is also recognized that tropical forest loss is having a direct effect on the build-up of greenhouse gases in the atmosphere, partly through the release of carbon dioxide (CO_2) as a result of burning, but also because trees are no longer available to fix CO_2 in the future. It has been estimated that some 1.8 gigatonnes of carbon are being released into the atmosphere every year as a result of tropical forest destruction, representing about 25 per cent of global anthropogenic carbon emissions[37]. In addition, clearing old growth temperate forest and replacing it with plantations usually results in a net carbon loss, in that most of the carbon is stored in the soil and humus layer, and much is released following clear-felling[38]. Although young plantations absorb high levels of carbon, much of this is released fairly soon after they are felled, so old-growth forests form the best long-term carbon sinks.

Invasion of pristine environments

In many countries, a more significant factor than the trees destroyed by logging is the opening up by the timber industry of previously pristine areas of forest to incursion by humans and invasive species. The commonest form of invasion is through the construction of logging roads. These have a number of significant environmental and social effects:

❑   as a direct form of damage, increasing erosion and flooding;
❑   as a migration route for weeds, introduced pests, diseases etc; and
❑   as a migration route for humans.

In Indonesia, construction of skid roads increased soil erosion from virtually zero to 12.9 imperial tons/ha/month[39]. In North America, road building also causes erosion and damage to water systems (and is the one kind of forestry damage readily admitted by the industry). Badly constructed roads can themselves result in major deforestation; in some areas 40 per cent of the forest may be cleared to create roads and tracks[40]. Roads also lead to forest fragmentation, and may isolate species in progressively smaller areas of habitat.

Logging roads also bring the risk of the spread of serious pest and disease outbreaks[41]. In the Kenai Peninsula in Alaska, the western spruce budworm

(*Choristoneura occidentalis*) is causing large-scale dieback in some areas. Although similar attacks have occurred in the past, it is suspected that human intervention has helped the pest to spread, particularly along logging roads and the tracks cut for seismic testing in oil exploration[42]. In New Zealand, the introduced Australian bush possum (*Trichosurus vulpecula*) has migrated along logging tracks into areas of natural forest, where it now plays havoc with indigenous wildlife.

Probably most important of all, roads can act as migration routes for illegal settlers, miners and hunters. The effects of the Trans-Amazon highway have been well documented, including mass deforestation of Rondônia state, development of new shanty towns and spread of poverty, as well as destruction of tribal communities. Building roads into previously inaccessible areas means that people follow in their path, and often start a cycle of clearance, felling and burning in what is often a fairly hopeless attempt to carry out permanent agriculture. Other users of the road include illegal miners in many tropical forests, along with political refugees and the landless looking for a stake. (This issue will be returned to in the section below on the importance of the timber trade in forest loss.) Such incursions also have direct effects on indigenous people; roads act as a corridor for diseases which can decimate populations who do not have natural immunity and resistance, while illegal settlers often provide a second serious threat. Road-sensitive wildlife species avoid areas with high road densities; for example, in North America bears and wolves are not generally found at road densities greater than one mile of road to one square mile of forest.

THE TIMBER TRADE AND NATURAL-FOREST LOGGING

The timber industry does not operate in the uncertain and often difficult conditions presented by logging natural forests without good reason. At the same time, there is no single reason for logging that applies in every part of the world. Among significant motivations are:

❑ cheap timber;
❑ artificially low costs through subsidies, cheap labour and special deals;
❑ as a stop-gap before plantations reach maturity;
❑ to clear away natural forests for plantation establishment;
❑ for special access as a result of corruption; and
❑ as part of an overall development strategy.

In many tropical countries concessions have, until recently, been cheap or free. In most temperate forest countries native-forest logging long ago progressed through a period of exploitation by private owners; most operations are now centred on remaining forests, held mainly by the state. Here, incentives for logging remain complex. Many national or regional governments actively support logging. In British Columbia, the Forestry Ministry has been losing large amounts of money through expenditure exceeding revenue; in 1989 excess costs reached US $450 million, thus costing British Columbia citizens about US $0.65 for every cubic metre logged[43]. The Indonesian government has supported logging and the development of plantations. In many countries government ministers are involved in logging, or have relatives or friends involved in logging, and thus turn a blind eye towards infringements of regulations, a situation that is currently allowing extremely rapid deforestation of Cambodia[44]. Elsewhere, powerful groups such as the army have been involved in natural-forest logging, thus maintaining exploitation systems which certainly do not benefit the majority of people. Military involvement in logging has been particularly important in Thailand during the 1980s.

In the Tongass National Forest in Alaska, one of the world's largest surviving temperate rainforests, development and logging have been encouraged by the state government. There were two reasons for this decision, to generate employment, and hence increase the population of the state; and because the Japanese government sought to buy Siberian timber to reconstruct the country after 1945 and the US government was anxious to provide an attractive alternative. In 1947, the Tongass Timber Act allowed long-term timber contracts to be signed within the National Forest, although there were still outstanding native claims on some of the area. Despite the offer of extremely attractive 'sweetheart deals', including 50-year leases, only two really large concessions have ever been signed, with Louisiana Pacific and the Japanese-based Alaska Pulp and Paper Company although the latter was cancelled in 1992. Other companies have toyed with the prospect and later rejected it, including the timber giant Georgia Pacific, which signed a deal and then later defaulted. The government has lost hundreds of millions of dollars in below-cost timber sales, and over half the timber cut is exported to Japan each year.

As the timber trade begins to feel the pressure of conservation legislation, forests are logged to secure land for future plantations. In Tasmania, Australia, the forest industry sees its future in the establishment of plantations of softwood, primarily *Pinus radiata*, and various eucalypts. There has been a recent increase in rate of felling of old-growth forests, for several reasons. First is an aggressive new marketing strategy – by state, national and transnational companies – aimed at expanding the export market to Japan. Secondly, and perhaps more important in the long term, is a desire to gain access to the soil under old-growth forests for the establishment of plantations; so-called 'soil liberation'. Linked to this, the industry fears that conservation legislation will put a large proportion of forests off-limits for logging, leading it to adopt a tacit policy of getting as much as possible in the interim. Production has been gradually accelerated since the 1970s, based around the pulp and paper and the wood-chip industries, and especially through larger exports to Japan. Tasmania now supplies about 40 per cent of Japan's hardwood wood-chip imports, through three woodchip export plants.

In some tropical forest countries, logging is seen as the first stage in the planned and deliberate clearance of a proportion of the forest estate, to provide farmland, space for hydroelectric development, road infrastructure and urban areas. Whereas this has happened in an *ad hoc* manner in countries such as Thailand, India and several West African states, it is being practised explicitly in, for example, Malaysia, Indonesia and some parts of Brazil.

How important is the timber trade in loss of native forests?

The timber industry often claims that its role in the loss of native forests is negligible when compared with other factors, particularly the impacts of farmers and settlers. This claim gained some weight when the FAO identified farming as the largest cause of deforestation in the tropics. However, further assessment shows that this analysis is geographically limited in its validity and that it ignores some important secondary effects of logging. Native-forest logging is important for two main reasons:

❑ logging is currently the major threat to surviving native forests in most of the temperate and boreal countries that still contain extensive tracts of natural forest, and in several tropical countries with a large forest estate, such as Indonesia, Zaire and the Central African Republic;

❑ logging is the primary incursion into other areas of forest, thus facilitating later in-migration, settlement and complete deforestation, particularly in tropical countries such as parts of Malaysia, and many West African and Latin American states.

A particularly contentious issue relates to the amount of tropical timber entering the rich countries of the North. The trade has long claimed that quantities were negligible, and in the mid-1980s used to ascribe the vast majority of exports to Japan. Japan certainly is responsible for a huge trade in tropical timber. However, several research projects over the past decade have shown consistently that large quantities of timber also enter Europe and North America. Indeed, if RWEs are used, the three large blocs (Japan, Europe and North America) have until recently accounted for most internationally traded tropical hardwoods and imported approximately equal amounts[45]. Today that situation is almost certainly changing slightly as other importers, including many tropical countries, grow in importance[46].

Illegal logging operations

One key element in the impact of logging on natural forests is the amount of illegal logging that takes place, particularly but not exclusively in the tropics. While a number of tropical countries can point with some pride to the area of their forests set aside for national parks and reserves, the reality is that many of these protected areas are little more than paper declarations and that illegal logging continues to degrade areas supposedly set aside for conservation. Timber theft has also been occurring on Forest Service land in the western US, and on private forest land in the UK.

Research into illegal logging is difficult, laborious and sometimes downright dangerous. Nonetheless, painstaking investigations over the past five or ten years have uncovered an enormous network of illegal trade around the world; a trade which in some cases dwarfs the legal trade reported in official statistics. The illegal trade has to date been studied most carefully in the Asia-Pacific, and reported by Debra Callister in a Traffic Network report *Illegal Tropical Timber Trade: Asia Pacific*. She identified six different types of illegal activity: illegal logging; timber smuggling; transfer pricing (discussed in Chapter 3); under-grading, under-measuring and undervaluing; misclassification of species; and illegal processing of timber. The scale of illegal logging is extremely hard to estimate with any accuracy, but all the signs are that it is massive. Virtually all logging for export currently taking place in India, Laos, Cambodia, Thailand and the Philippines is illegal. Estimates suggest that a third or more of Malaysian logging may be illegal and within Indonesia up to 95 per cent, at least, is not *wholly* legal[47].

Illegal operations are not confined to Asia. Research suggests that as little as 10 per cent of timber logged in Brazil is exported legally. Much of the broadleaved mahogany (*Swietenia macrophylla*), mostly for sale to the UK and the USA, comes illegally from Amerindian reserves in Rondônia and elsewhere. In the 1980s trade grew by 370 per cent. In 1990, 56 per cent of exported mahogany went to the UK. An investigation in 1992 found that five companies in the UK and the USA were unwittingly importing mahogany cut illegally in the Uru Eu Wau Wau reserve in Pará state, where it is estimated that logging for mahogany has resulted in the opening up of 250,000 km^2 of primary rainforest through construction of logging roads and additional damage to other tree species. Settlers and miners follow the logging roads and cause further deforestation. At least 92 indigenous peoples' groups are affected by mahogany extraction, despite clear legal protection of their lands, and violent clashes have resulted. José Lutzenberger, former Brazilian Secretary for the Environment, has admitted the impossibility of addressing the illegal timber trade in Brazil and said: 'As there is little we can do to stop the supply, it is up to the people of Britain and other First World countries to stop the demand'. In 1993 injunctions were imposed, and eventually upheld, in Brazil against three companies – Perachi, Maginco and Impar – forbidding them to log in Amerindian reserves in Pará. The companies were among 22 on a list of approved suppliers from the British Timber Trade Federation, which has signed declarations that members do not trade in illegally logged wood[48]. Other species of Brazilian timber are also felled illegally. Brazilian rosewood (*Dalbergia nigra*) is highly sought for making musical instruments. It is now so scarce that it was accorded the rare honour of a global ban on trade in 1992, but an illegal trade is still continuing[49].

The illegal trade is alive and flourishing in Africa as well. Aid money, poured into Ghana as part of a Structural Adjustment Programme attempt to boost the timber trade, led to illegal forestry practice on a massive scale. Several European agents, exporters, merchants and suppliers consorted with local companies in illegal operations. At least 11 of the 13 'foreign' companies that benefited from British aid and credit in the supposed revitalization of the timber industry have been implicated in fraud or other malpractices and out of 21 'foreign' companies receiving loans from the World Bank, so far 15 have also been investigated, implicated or found guilty of irregular or fraudulent practices. According to calculations by Friends of the Earth International, the result was felling of 250 km^2 of rainforest in the country. By August 1990, 106 firms and individuals had voluntarily repatriated a total foreign exchange of £6.6 million, although it is estimated that at least £30 million has been lost to the Ghanaian economy[50].

In Kenya, logging of indigenous forests has been banned except for areas of private land and in selected forest reserves, but a brisk trade in indigenous timber continues. Ways to 'legalize' stolen trees include: illegal felling and pit sawing (often at night); ring-barking trees so that they die and can then legally be extracted; bribery; duplication and re-use of felling licences;and use of forged documents[51].

NATIVE-FOREST LOGGING AROUND THE WORLD

Native-forest logging is most significant, in terms of area cleared, in those areas where substantial amounts of primary or old-growth forest continue to exist. However, from a biodiversity perspective, equal or greater importance can be attached to remaining natural or semi-natural fragments in areas, such as Europe, where virtually all forests have been cleared and managed continually for hundreds or thousands of years.

In broad terms, conditions in Pacific Rim forests can be distinguished from those in the Atlantic forests. The Pacific Rim countries, including the western American coasts, Australasia and the Pacific Islands, eastern areas of the Russian Federation (in fact, the analysis applies to most of Siberia) and Southeast Asia still contain substantial areas of primary forest, despite a marked acceleration of logging in recent decades. In the Atlantic countries, including Europe, the eastern coast of North America and temperate coastal areas of South America, and some West African countries, virtually all primary forests have long ago been felled. This distinction is approximate; some Atlantic African countries contain large remaining forest estates as does Brazil and some Central American states. But it does give a broad picture of the geographical limits of the native-forest logging debate. Although a book of this size cannot attempt to give a complete overview of the impacts of logging on native forests, a summary is presented in Table 4.2 and some case studies are briefly outlined below.

Table 4.2 Native forest logging around the world

Country	Status and details
Europe	
Finland	Only 1–2% old-growth forest remains; this is still being logged in places.
Latvia	Logging has increased 700% in the last few years, mainly for the export market, threatening many important wet forests.
Norway	Logging of remaining old-growth forest has increased since plans for additional conservation legislation were suggested.
Poland	Logging has intensified since 1989, and is taking place on the edge of the internationally important Bialowieza forest Biosphere reserve.
Sweden	Logging of old-growth forest continues in the boreal region, despite being reduced to 1–2% of the original.
UK	Illegal felling of broadleaved trees to sell as firewood is on the increase.
Russian Federation	Logging is occurring in many biologically rich areas of Siberia and European Karelia. In the latter case there is currently a growing cross-border trade in birch with Scandinavia.
North America	
Canada	Boreal forest logging is taking place on a large scale in many areas, including particularly Alberta. In Ontario, two-thirds of the remaining 1% of old-growth forest is slated for commercial felling.
USA	Logging of old-growth forests in the Pacific Northwest looks likely to increase again in response to Republican aims to deregulate the industry and overturn environmental legislation.
South America	
Argentina	Temperate forests are rapidly being logged by foreign companies, including many from North America.
Bolivia	Forest loss has now reached critical levels in some areas.
Brazil	Illegal logging of mahogany is having a major impact on the ecology, and the survival, of forests in many areas, and until recently 80% of mahogany exports were of illegally felled trees.
Chile	Large areas of beech (*Nothofagus*) have been logged to make way for pine plantations in the last decade, often by foreign companies, and *Araucaria* forest is also threatened.

Table 4.2 continued

Country	Status and details
Guyana	Increased logging by foreign companies is now threatening one of the largest remaining areas of pristine rainforest in the region.
Suriname	Malaysian, Indonesian and Chinese companies are preparing to log in pristine rainforest.

Africa

Cameroon	Numerous transnational companies are operating in the country, including companies from Belgium, France, Germany and Italy. A survey in 1993 identified 100 forest operations, 60 of which were foreign-owned. Logging has increased 100% in the last few years.
Congo	At least 15 of 36 active timber companies are foreign-owned, controlling about half the cut and based in Germany, the Netherlands and France.
Côte d'Ivoire	Less than 14% of the original forest remains. Companies from Denmark, France, Germany, Italy and Holland remain active.
Gabon	Most timber production is under European control, predominantly from France but also from Germany, Italy and Switzerland. Latest estimates for deforestation are 0.6% per year.
Ghana	More than 90% of forests have been logged since the 1940s. Danish and Dutch companies operate, and in the late 1980s a state-owned timber company was rehabilitated by a UK company; this was abandoned after allegations of corruption.
Nigeria	Much of Nigeria's small area of remaining forest is threatened by legal and illegal timber operations.
Zaire	Around ten timber companies are operating in Zaire, and most logging is carried out by foreign-based firms from Belgium, Canada, Denmark, France, Germany and Italy. Logging is increasing rapidly.

Asia

Cambodia	Illegal timber felling has increased enormously over the past few years and is rapidly depleting the country's forests.
Indonesia	The government intends to replace 2 million ha of forest with plantations by 2000. Commercial forestry is a major cause of forest loss in Kalimantan, Irian Jaya and other outer islands such as Siberut.

Table 4.2 continued

Country	Status and details
Laos	Illegal logging has increased rapidly as a result of a ready market created in Thailand due to the latter's logging ban.
Malaysia	Logging is the major cause of forest loss in Sabah and Sarawak, and is still important in some areas of Peninsular Malaysia.
Philippines	Logging has already caused major deforestation in the country. Illegal logging is now more important than legal operations and is still a major source of exports.
Thailand	Illegal logging continues despite a logging ban, particularly in the north east and on the Burmese border.
Vietnam	Large areas of the country are being cleared of natural bamboo to feed pulp mills.
Pacific	
Australia	Logging is the major cause of forest degradation and loss, particularly in the south west and Tasmania.
Papua New Guinea	Logging, including illegal logging, is the major cause of forest loss in PNG, mainly involving expatriate firms from south east Asia.
Solomon Islands	Legal and illegal logging is the major cause of forest loss.
Vanuatu	Logging is increasing rapidly, mainly controlled by expatriate Malaysian companies.

CASE STUDY: THE RUSSIAN FEDERATION

The Russian Federation contains the largest remaining area of temperate forests; almost 1 billion ha or 20 per cent of total global forests. Siberia is said to contain over 60 per cent of the world's remaining boreal old-growth forest[52]. The forest is habitat for many rare wildlife species. At least 26 indigenous groups live in the Russian forests[53]. Russian forests contain what is almost certainly the largest remaining area of old-growth forest in the world, and are thus discussed in some detail below.

The forest industry

The forest industry has long been important within the country. Volumes peaked at 395 million m^3 in 1975[54]. Although much of the timber is used internally, exports were worth US $ 3.75 billion in 1989[55]. Forest protection, while being the subject of extravagant claims in the past, remains low.

Environmental problems

As early as 1978, Boris Komaroz (pseudonym) accused the USSR of poor forest management and high losses from forest fires[56]. Analysts also claimed that the most accessible European forests were being overcut[57]. The collapse of the communist government has brought about a period of serious economic and social unrest. In the Russian Federation, poverty and malnutrition are now widespread and the living conditions of most people have deteriorated over the past decade. There has been a breakdown in central government control, a rise in organized crime and widespread corruption among local government officials. The volatile situation has created a number of effects within forests:

❑ a 30 to 50 per cent decrease in logging by the state-owned *lespromkhos* timber companies since 1990;
❑ an increase in exports, often accomplished by bribery, since exports can raise 1000 per cent more than domestic sales;
❑ a resource crisis for industries within the Russian Federation, some of which have had to cease production because of timber shortages, with all 300 domestic furniture and timber processing companies suffering a dramatic decline in production[58];
❑ a number of short-term investments by small foreign companies, interested in making quick money while the situation remains unstable and unregulated;
❑ a generally more cautious, but potentially very large-scale, interest from large Western TNCs wishing to buy concessions within Siberia;
❑ more active involvement from some of the Southeast Asian timber companies especially from South Korea and Japan; and
❑ stringent economic demands, including structural adjustment, from the World Bank which encourage rapid resource utilization and hence the increased rate of forest felling[59].

Logging has increased in some areas[60]. Most forest is clearcut, by both state-owned companies and TNCs. The forestry industry is wasteful – about 20 per cent of the roundwood is left behind at the logging sites, and overall losses are said to be between 40 and 70 per cent[61]. There are serious problems of regeneration of clearcut forest in some boreal regions and also a tendency for multiple-species forests to be replaced by predominantly monoculture forests. River transport of large amounts of clearcut logs damages river systems and results in serious wastage as many logs sink. There is strong evidence that harvest levels are being exceeded in many places[62]. Most local people are not gaining from forestry operations and profits are going out of the country or into the pockets of corrupt officials. Although a few of the native groups have organized, most are all but powerless[63].

Although effects on wildlife are still poorly understood, reports of problems are starting to appear. The impact of logging on the habitat of the Siberian tiger (*Panthera tigris altaica*) was recently exposed by *National Geographic* magazine which claimed: 'clearcutting and poaching are twin gun barrels pointed at the Siberian tiger'[64]. A report on biodiversity in the Russian Federation, prepared by WWF in 1993, identified logging as a problem for many forests, including in the boreal region, the Kola-Karelian region, the Ural Mountains (where 75 per cent of forests have been destroyed), Zabaikal, Amur-Sakhalin, and throughout Siberia[65]. According to the deputy chairman of the Federal Forest Service:

> *Over-exploitation of forest resources, violations of ecological and forestry regulations and poor forest management during the last decades has drastically depleted forest resources; if timber continues to be logged at the present rate, assuming there is no waste in timber processing, climax forests will be completely destroyed in 40–60 years[66].*

The timber trade moves in: European Russia

Logging has traditionally taken place in areas accessible to road or rail transport. European Russia holds 20 to 30 per cent of the country's total growing stock, but more than 60 per cent of the annual cut takes place in the area. Over the last few years, it is estimated that the annual cut in European Russia has been overcut by some 700 million m^3, an amount equal to the total of five years' annual allowable cut for the area[67]. Wood products are the principal export from Karelia, and Finland is the main export market for Karelian timber[68].

In 1987, Land and Timber Services International began to explore market and trading possibilities within the Russian timber industry. Three sawmill projects were identified for joint investment; one based near the Black Sea and two in the Ural region. The project has included the installation of new machinery and training from western staff[69]. Since then, a range of initiatives has been developed or extended, including:

❑ A Russian-Finnish joint venture at the Chudovo-RWS birch plywood plant near St Petersburg which reached a production rate of 50,000 m^3 in 1992. Birch logs are supplied by Novgorodlesprom, formerly a state company, which holds rights to raw material in the area[70].

❑ Tramed, a Russian-American joint venture, is planning a modern planing and stress-grading mill in Karelia[71].

❑ Finnish companies have increased their interest in the area, and in at least one case have opened up old concessions that were agreed before the 1917 revolution and have laid dormant ever since[72].

❑ The Finnish company Enso-Gutzeit Oy has gained rights to log in Ladoga over 380,000 ha, of which 50,000 ha is a nature reserve in the form of a 2 km strip along the coast. Despite thinning rather than clear-felling, the company has been accused of causing damage to young trees and was in 1991 fined over 2 million roubles as a result[73].

❑ Another Finnish company, Tehdaspuu Oy, is involved in a joint venture in Tepules to fell timber on the islands and shores of Karelian lakes, in violation of watershed protection regulations. In November 1991, the Public Prosecutor of Karelia instituted criminal proceedings against the company for illegal logging, felling of young trees etc, in part due to prior permission having been given, without proper authority, by another Karelian official[74].

❑ Forest Starma, a Russo-Norwegian joint venture, operating in the Amur region: 5.5 million m³/year is predicted, 40 per cent to foreign partners. Norwegians delivered felling equipment and temporary housing[75].

❑ Skogsallians, a Swedish timber company, has formed a joint venture in Russia. Most of the timber is claimed to come from thinnings, with only 10 per cent from clearcuts. The company states that it will meet full Swedish environmental standards and that NGOs are welcome to visit the area and inspect holdings[76]. The company will operate over a wide area east of St Petersburg and Moscow.

Timber trade: Siberia and the Far East

Siberia accounted for 45 per cent of the former USSR's overall timber resources[77]. West Siberia is 60 per cent forested, East Siberia 70 per cent forested and the Far East 33.5 per cent forested[78]. Owing to the lack of roads and railways much of the Siberian forest is inaccessible. Logging is therefore concentrated in the Siberian Far East and in corridors along the railways of Central and Western Siberia. The primary lumbering centres in the region are the Tyumen, Tomsk, Krasnoyarsk and Irkutsk regions[79]. In East Siberia, good forest resources, together with the availability of hydroelectric power, have led to the creation of a large forestry complex at Bratsk-Ust-Ilimsk in Irkutsk Oblast[80].

Amidst a rapidly growing interest in Siberian timber, a few foreign countries are coming to dominate the debate and practice of logging, particularly in the boreal regions. These include Japan, North and South Korea and the USA.

Japan

Japan had long imported timber from the USSR; the break-up of the Soviet Union and the emergence of a free market in the Russian Federation will probably lead to increased trade. The USSR was the source of Japan's cheapest timber, mainly from Siberia. Major port facilities were planned in the region, although most have not been built. Japan has recently arranged to import hardwood chips as well. It has encouraged a barter system with the Russian Federation, exchanging equipment for logs. It has been suggested that Japanese firms may be reselling Russian timber to South Korean companies, thus involving them in this aspect of trade[81].

Several Japanese trading houses, including C Itoh, Mitsubishi, Nissho Iwai and Marubeni are already active in the Russian Far East[82]. Foreign involvement also includes technical expertise. In 1992, Japan's KS Industries, a consortium of ten Japanese trading and heavy machinery companies, agreed to exchange US $700 million worth of lumbering machinery, sawmill production equipment and heavy equipment for building roads for 6 million m^3 of raw timber and 400,000 m^3 of processed wood. Other trade deals involve the Japanese Tairiku Trading Company and the Russian Federation's Severovostokzoloto, a joint venture to develop natural resources in the Far East[83]. Among the Japanese companies already active in the country, the first Japanese joint-venture sawmill opened at Lidoga in the Russian Far East[84]. A similar deal exists between a Tokyo-based venture of 12 Japanese trading houses and the east Siberian government of Buryat to develop forest resources, where timber will be exchanged for medical equipment, food and consumer goods[85].

North Korea

Two joint Russian-North Korean timber companies are logging tracts of Siberian forests, using the forced labour of between 15,000 and 20,000 North Korean prisoners. The companies are Urgalles, which logs along the Urgal-Izvestkovaya railway in the Khabarovsk region of Eastern Siberia, and Tyndales which operates in six or seven areas around Tynda in the Amur region.

The enterprise originated in the 1960s and the agreement was extended for a further three years in August 1991[86]. North Korea and the Russian territory of Khabarovsk extended their timber harvest joint venture for another ten years[87].

South Korea

South Korean firms have been involved in logging, including illegal operations in the mountainous regions of Shoris and Khabarovski in the far east of the country, despite the fact that these areas were declared nature reserves in 1990[88]. The giant *chaebol* Hyundai, through its Svetlaya joint venture, is developing resources in the Russian Far East in cooperation with Japan's Mitsubishi company and exporting timber to Japan through the M and H Corporation. In October 1992 the Greenpeace flagship *Rainbow Warrior* blockaded a barge owned by Hyundai which was being loaded with illegally clearcut logs. Greenpeace claimed that Hyundai and its Russian partner had broken an agreement with the regional authorities to cut only dead and dying trees, and had instead been systematically clearcutting the area. Agreements for 'reforestation' of the logged area had also been unfulfilled. The Hyundai company is interested in logging the Bikin River watershed, to cut 1 million m³ of timber a year for 30 years. Siltation as a result of clearcutting has affected the local fishing industry and drinking water supplies have been polluted through oil spills from Hyundai machinery[89]. The Russian government has recommended that the land be protected and handed to the control of the Udege people – between 2000 and 3000 live in the area – but despite a Russian Supreme Court decision supporting a veto, the regional administration is apparently continuing to take steps to allow logging[90].

USA

The US TNC Weyerhaeuser negotiated for two years to set up a joint venture along the coast of the Khabarovsk region in the Russian Far East, which would open up 1 million ha of forest land for exploitation. This includes extensive old-growth forest in the Botcha region, which is important ecologically and as the home of the Orochi, an indigenous group of about 300 people who live by fishing the offshore waters. Weyerhaeuser launched a large public relations exercise in the area, taking local officials to the USA to see model tree farms and putting money into local scientific institutes, some of which had opposed the development. However, opposition to the scheme remains strong in the area[91].

Other US companies are known to be interested in the area: Georgia Pacific and Lousiana Pacific have both entered into negotiations at various times, but without committing themselves to any logging[92]. Poor accessibility and infrastructure, regeneration problems, low returns and the risks of importing alien pest species have previously deterred some US companies from operating in the region[93]. The USA banned import of raw logs from Siberia in 1990 because of fears of introducing disease, and a Russian-American-Norwegian joint venture is apparently planning to build 11 irradiation plants in the region to treat logs for export[94]. A Memorandum of Understanding was signed between the Russian Federation and the USA in June 1994 to facilitate a greater timber trade between the two countries, initially concentrating on timber from 500,000 ha of Siberia[95].

Other foreign investments in the area

Several Taiwanese companies are reported to be exporting timber harvesting equipment to the Russian Far East in exchange for timber products. The Norwegian enterprise Forest Machine Technology has been logging in the Khabarovsk region since 1992, exporting wood chips and logs to Japan. After one failed attempt to start logging, the current operation involves a Russian partner in a joint venture at Vanini on the Russian Pacific coast. The

Norwegian government has supported the project with a grant. Although the company claims it will practise 'sustainable forestry' no details of what this means have been given[96].

Government proposals

There are, in theory, intentions to introduce further forest conservation measures. The Russian authorities have planned to enlarge 'special forests', where cutting is restricted, from 24 to 27 per cent of the total, and ecologically fragile areas, where cutting is prohibited, from 8 to 12 per cent[97]. In June 1991, a Russian forestry journal stated:

> *The logging situation in the Russian state has become critical. At the start of 1991, the planned cut is put at 67 million m³ less than the previous year's cut; that is, a drop of 17 per cent from 394 million m³ to 327 million m³. By increasing areas of protection forest, including those along fish-spawning rivers, about 24 billion m³ out of a total of 71 billion m³ have been withdrawn from exploitation[98].*

A new Forestry Act was adopted by the government on 6 March 1993[99]. Recently, Tripartite Instructions have been issued to levy a new 20 per cent tax on commercial logging enterprises. Revenue is intended to support the maintenance and protection of forestry resource[100]. The Department of Forest, Pulp and Paper Working Industries has been established at state level under the Ministry of Industries to supervise the approximately 3000 enterprises engaged in forest-based industries[101]. In early 1993 a state-run timber industry, Roslesprom, was set up to replace the Soviet Ministry of Forestry, to regulate the sector's investment, scientific-technical and export activities; it hired only about 10 to 12 per cent of the former Soviet Ministry of Forestry's workforce[102]. In 1994, a development programme was announced to increase cutting from 120 to 140 million m³ to 300 to 350 million m³/year[103].

Conservation organizations in the Russian Federation believe that there are serious problems with the Forestry Act as it stands and that the pressure will be to cut rather than to conserve. Public participation has been reduced with respect to decision making. Significant powers have been devolved to local authorities, which are often heavily influenced by criminal elements and corruption[104]. There are also serious fears that the reserves will not be adhered to in the present political chaos. Ecologists warn that nowhere near enough forest is preserved, and that the planned clear-felling operations could have serious ecological side effects.

The breakdown of the former USSR has created an institutional vacuum, in which foreign and domestic companies, sometimes acting hand in hand with corrupt local officials, are having a rapid and in some cases devastating effect on natural forests. The new Forestry Act fails to address these issues adequately, and in any case is likely to be effectively ignored over wide areas of Siberia. The World Bank is encouraging the process of rapid resource exploitation through its own policies. It is likely that areas of Siberia will be among of the most important battlegrounds about natural-forest logging in the next few years.

CASE STUDY: CANADA

Canada contains about 6 per cent of the world's forest resources, totalling some 226 million ha. This is almost twice the total forest area of Europe, excluding the Russian Federation. Forestry contributed Can $26.7 billion to export earnings in 1993, up 16 per cent on the previous year[105].

There is a fierce debate about forest management in Canada. The Canadian government argues that on a national basis forestry is practised sustainably[106]. However, this claim has been repeatedly challenged by environmental groups, both in terms of maintaining cut and with respect to the environmental sustainability of current practices[107]. The exceptionally low stumpage rates paid in Canada have resulted in claims that government is in the pocket of big business, and is using taxes to subsidize company profits.

Conservationists question how Canada can lobby against tropical rainforest destruction while allowing the destruction of its own rainforest. The links between rainforest in Canada and Brazil have been illustrated in a paper from a government research department[108] and some activists have dubbed Canada the 'Brazil of the North'. This provokes strong reactions from industry representatives, who claim that the government was duped into publishing a paper from an environmental extremist and that comparisons with Brazil are irrelevant, in that Canadian forests regenerate after felling while the Amazon is being replaced with ranches and pasture land[109]. Until recently, most international attention has focused on British Columbia, although forest issues are also a priority in several other provinces.

British Columbia

Nowhere in Canada is the debate about forest policy more polarized than in British Columbia. Despite a long history of timber harvest, British Columbia entered the second half of the twentieth century with much of its original primary forest intact, mainly in areas of the provincial-owned forest. In 1991, British Columbia accounted for 62 per cent of the softwood lumber, 29 per cent of the pulp, 16 per cent of the paper and 84 per cent of the softwood plywood manufactured in Canada[110].

However, during the 1970s and 1980s there was a substantial increase in the rate of cut. Analysis by the Sierra Club of Western Canada estimated that over half of all timber cut in British Columbia since 1911 has been since 1975, and that about 1000 square miles (260,000 ha) of forest is felled every year. The annual cut within the province now exceeds that from all the USA's national forests combined. The sustained yield is estimated at 59 million m³/year, whereas the 1989–90 cut reached 78 million m³/year[111]. Much of the province's most productive ancient forests have been cut in huge swathes along 500 miles of the coast. Whilst most cuts take place away from main roads, and are thus often missed by the traveller on the ground, a patchwork of vast clearcuts can be seen over hundreds of miles by anyone flying along the west coast.

An approaching forest crisis

At the start of the 1990s, the rapid rate of felling looked set to continue. Plans to log new areas provoked strong opposition within the province. Two Tree Farm Licenses on Vancouver Island, which contained much of the island's remaining old-growth forest and covered the equivalent of one fifth of the island, were listed for felling by MacMillan Bloedel, a company based within the province, and by the New Zealand owned Fletcher Challenge group[112].

Of particular concern to environmentalists was the small area of protected forest. In 1992, protected land covered just 5.3 per cent of the province, of which only about two per cent was temperate rainforest. In Vancouver Island, only six of the 89 largest watersheds remain unlogged[113]. A detailed inventory of protected watersheds in coastal temperate forest identified nine entirely protected watersheds of more than 5000 ha, with only one of these exceeding 20,000 ha[114]. Thirteen other similar watersheds were partially protected, along with 22 smaller watersheds between 1000 and 5000 ha. This total means that only 6 out of 17 ecosections in the coastal temperate rainforests in coastal British Columbia contain representative, protected and entire primary watersheds larger than 1000 ha[115]. Yet the provincial government has committed itself to protecting a representative selection of all 100 ecosections in British Columbia by the year 2000.

Felling is also causing environmental damage. Most logging is through clearcuts[116]. The industry argues that clearcutting is 'often the most appropriate way recognized by professional foresters to harvest, salvage and renew most types of forests in Canada'[117]. However, soil erosion is admitted to be a serious problem in clearcut areas. A 1988 Forest Resource Development Agreement report estimated annual economic loss to the British Columbia economy through soil erosion at $80 million/year, and that 'The annual [wood] productivity loss resulting from soil degradation is estimated to be increasing by 50,000 m³ or $10 million per year. If present forestry practices continue, the annual loss may double in less than ten years'[118]. The erosion has knock-on effects for water purity, fish breeding and other wildlife. Pollution from pulp mills adds further environmental problems.

A declining economy and loss of control

Despite the rise in cut, logging has been providing less money, and fewer jobs, within British Columbia. Between 1979 and 1989 actual cut grew from 77 to 85 million m^3, whereas direct forest jobs fell from 96,000 to 88,000 as mechanization pushed loggers out of work. MacMillan Bloedel has laid off workers and is apparently looking for more investment opportunities outside the province[119]. The Forestry Ministry lost money as well; expenditures exceeded revenues by Can $450 million in 1989, thus reportedly costing British Columbia citizens some Can $0.65 for every cubic metre logged[120].

Although most of the land being logged is within the province's ownership, control of the forest companies carrying out the logging lies mainly outside British Columbia, and some 36 per cent is in foreign ownership (an issue addressed in Chapter 3). Among foreign companies holding concessions or timber and paper mills in British Columbia are: Fletcher Challenge from New Zealand; Eurocan, which is a joint Canadian-Finnish company; and Weyerhauser from the USA; there are also Japanese interests.

Recent developments

International criticism grew rapidly during the 1990s, with Greenpeace mounting a high profile campaign including direct actions against a number of Canadian embassies around the world. Within the province, thousands of protestors focused their attention on felling in Clayoquot Sound, a particularly disputed 262,000 ha area to the west of Vancouver Island, and over 800 were arrested in 1993. Criticism initially produced a sharp reaction from within the country. The business community defended the cut, and a 1991 editorial in the *Financial Post* accused conservationists of 'phoney charges', saying that clearcuts made sense and that: 'Many mature stands of old-growth forests are as much monocultures as what replaces them after they are cut'[121]. Business also funded an international public relations company, Burson-Marsteller, to improve its image. As a result, the 'BC Forest Alliance' was launched; this is supposedly a citizens' group to defend forestry in Canada, and employs a former Greenpeace member as a consultant, but is funded by the industry. The Alliance made attempts to split opinion against conservation groups and to head off further criticism[122].

However, pressure from inside and outside Canada started to have an impact on decisions made within the province. Newspaper articles and reports in Europe and the USA received a lot of media attention within the province. Threats of a boycott of Canadian timber, which were made by a number of groups within Europe, produced a strong reaction. In addition, the new provincial government promised to address forest questions.

The *BC Code of Forest Practice* was launched with the aim of improving forestry methods. It includes a reduction in maximum clearcut area, provision for protection of old-growth forests and wildlife, as well as limits to road building. The code has been criticized by environmentalists for not being tough enough[123]; on the other hand the forest industry claims that it is making too many demands and will seriously reduce profitability of the trade. There is evidence that the Code's standards are frequently ignored or broken in practice, and a recent study suggested that all nine companies operating in the area of coastal British Columbia had failed to meet guidelines on occasion, with compliance rate ranging from 50 to 80 per cent[124].

Various attempts to address the particular problems of Clayoquot Sound were made, including a gap analysis by World Wildlife Fund Canada, which identified the need for additional protected areas in the region[125]. Additional areas of Clayoquot Sound were set aside in April 1993, with 33 per cent fully protected, 17 per cent partially protected with limited logging and 45 per cent open to resource use including logging[126].

In the province as a whole, the number of national parks and conservation areas was substantially increased. Prior to 1992, under 6 per cent of British Columbia was protected, but by October 1994 some 81 new parks and protected areas had been created, raising the total protected area to 8.6 per cent, or 82,000 km^2 including the Kitlope River Valley, the world's largest remaining untouched coastal temperate rainforest[127]. By mid 1995, over 100 new areas had been created.

Perhaps most importantly, the provincial government established a *Scientific Panel for Sustainable Forest Practices in Clayoquot Sound*, with terms of reference to review and improve existing management plans, recommend research priorities and identify key ecological indicators. The Panel consisted of foresters, ecologists, earth scientists and representatives of Native American groups in the area. It issued a series of reports in 1995, looking at first nations' perspectives[128] and drawing up a vision[129] for forestry in the area. A major report on sustainable management recommends some radical changes including a *variable retention silvicultural system* which would include 'the permanent retention of forest "structures" or habitat elements (eg large decadent trees, or groups of trees, snags, and downed wood) from the original stand that provide habitat for forest biota; [and] a range of retention levels'[130].

The report recognises that the watershed is the basic unit for planning and management, seeks to approximate the natural disturbance regimes of the area and proposes that felling be modified to protect rivers and streams, wildlife areas and long term sustainability of timber harvest. It concludes that 'Clayoquot Sound is an excellent place to test the concept of local responsibility for sustainable ecosystem management'[131].

The Panel's report appears to offer the most coherent way forward for a beleaguered government, and for long term management of the forest resource. In July 1995, the provincial government accepted the Panel's recommendations and agreed to fully implement its 120 or more recommendations[132]. Considerable progress has certainly been made since 1990. However, in mid 1995 the provincial government was trailing in the opinion polls, and was opposed by a party with far less interest in placating environmentalists. If the government's commitments are to be carried through, continued vigilance by conservation organisations will be needed.

Other Canadian issues

Although forest issues in British Columbia have dominated the world headlines, other parts of Canada are also the scene of conflict with regard to forest use. In Ontario, only about one per cent of old-growth forest is left. Two thirds of this has been set aside for felling and, under current plans, will have disappeared by 2005. Of 100,000 ha old-growth red and white pine remaining, approximately 40,000 ha is protected, 20,000 ha is temporarily protected but may be harvested later, and over 30,000 ha is now open to harvest. The forest which regenerates after logging is radically different from the original ecosystem, changing from a landscape dominated by conifers to one with far more broadleaved species. Red and white pine both fail to regenerate effectively, meaning that the highest value timber is not returning after the first cut[133]. The resource is being mined rather than managed.

The role of Japanese multinational companies in opening up the vast boreal forests of Alberta has also recently come under scrutiny from environmental groups and politicians. Two out of six pulp projects started in the last five years originated with Japanese investment. The use of provincial funds to support Mitsubishi's Al Pac plant[134] and the clash between Daishowa and the Lubicon Lake people over its Peace River mill[135] have polarized public opinion. Both mills opened in the 1990s, and created a storm of protest about their environmental impacts and their long term social implications and economic viability. The two operators have been granted logging rights on 11.34 million ha, some 17.5 per cent of Alberta[136]. Daishowa's dispute with the Lubicon has continued for almost a decade and the Lubicon have accused the company of breaking verbal agreements not to log territory until a land dispute was settled[137].

The province has recently announced the development of a *Forest Conservation Strategy*, but environmental groups are wary of these plans and concerned about the political bias of those appointed to its steering committee[138].

Canada remains in an ambivalent position with regard to forest developments. The industry is amongst the most sensitive in the world regarding criticism from both inside and, particularly, outside the country. On the other hand, Canada is clearly aiming to play a leading role in setting international forest policy in the future, and has been responsible for several of the initiatives following the Earth Summit, as described in Chapter 7. Canada also caused a minor furore by proposing a hurriedly drawn up timber certification option to the International Standards Organisation in June 1995, which was considered to be too weak and finally withdrawn after sustained lobbying by environmental NGOs. However, despite staunch resistance by sections of the industry, changes are clearly coming within the country. The issue of jobs and the environment, which has split communities in the past, is now starting to be addressed, in part because it is clear that decline in forestry employment has more to do with mechanization than set asides. Overall level of direct employment has been falling for many years and for example between 1989 and 1992 some 60,000 out of a total of 348,000 jobs were lost in the industry[139].

The amount of old-growth forest that survives into the next century and beyond will be decided, to a large extent, by negotiation between the industry and others over the coming five to ten years. Attempts to increase protected areas, such as the WWF Endangered Spaces Campaign, are at present working alongside efforts to change forestry practices.

CASE STUDY: ALASKA, USA

Approximately one third of Alaska is forested. In the southeastern part of the state there are extensive areas of coastal, old-growth rainforest dominated by Western hemlock (*Tsuga heterophylla*) and Sitka spruce. Other common conifer species include Western red cedar (*Thuja plicata*), Alaska cedar (*Chamaecyparis nootkatensis*), and lodgepole pine or shore pine (*Pinus contorta*). In the interior, forest is dominated by white spruce (*Picea glauca*) and black spruce (*Picea mariana*), along with paper birch (*Betula papyrifera*) and some willow and aspen[140].

Alaska's forests provide habitat for some of North America's most threatened species, including the northern goshawk, bald eagle (*Haliaeetus leucocephalus*), grizzly bear (*Ursus horribilis*) Alexander Archipelago wolf (*Canis lupus*) and many deer species, such as the Sitka black-tailed deer (*Odocoileus hemionus sittensis*) which relies on old-growth stands to provide shelter during the winter.

Ownership of forests is divided between the government, in two national forests – the Tongass and Chugach – and various other areas controlled by the Bureau of Land Management; the State of Alaska Division of Forestry; and private owners, including especially various Native American corporations. Inuit people have had substantial areas of forest returned to them under the Alaska Native Claims Settlement Act[141]. Old-growth supports subsistence species, including fish, deer and bears, which are harvested by First Peoples. There are 111 national or state parks in Alaska. In 1980, the Alaska Lands Act set aside approximately 104 million acres, of which 56 million are designated as Wilderness. While by no means all of this is forest, there are many millions of acres of protected forest within the state. However, not all forest types have the same level of protection. The Tongass National Forest classifies old-growth by volume classes. The highest volume class, (7), containing the largest trees and oldest forests, is usually located at low elevations and is critical for deer, salmon and bears. Less than ten per cent high volume old-growth remains, the rest having been high-graded for felling[142]. In particular, high volume stands have been prioritized, and now only a fraction remains.

Currently, the annual cut of Alaska's forests is greater than the entire cut for the second decade of the century. Although exports have slumped from a 1980 peak, they remain at about 450 million board feet a year. Ninety per cent of exports go to Pacific Rim countries and 70 per cent to Japan alone. There, demand is mainly for raw logs, which fetch the highest prices. Under Alaskan state law – the Primary Manufacture Rule – timber taken from national forests must have undergone at least minimal processing in a local mill before export. This rule always applies to spruce and hemlock, but can be waived for cedar under some circumstances. Timber from private lands, including those owned by Native American corporations, can be exported in any form, and virtually all is sold as raw logs[143]. There are nonetheless more than twenty pulp and timber mills still operating in Alaska, five being run by Louisiana Pacific[144].

Despite its political profile, forestry employs less than 3000 people in the state. The amount of timber processed within the state is a key factor in employment figures, because research has shown that a far greater proportion of jobs are available in secondary processes than in logging[145].

The Tongass: one of the world's largest temperate rain-forests

The most critical forest conservation issue within Alaska is the continued exploitation of the Tongass National Forest. The Tongass extends for 16.9 million acres north of the Canadian border, covering 80 per cent of southeast Alaska, with 57 per cent being forested, the rest consisting mainly of mountain and tundra. It contains the world's biggest remaining populations of grizzly bear and bald eagle.

Extraction has been dominated by two large timber companies, the TNC Louisiana Pacific in the USA and the Tokyo-based Alaska Pulp and Paper Company, although the latter's contract has recently been cancelled by the US Forestry Service. These companies have been operating under extremely attractive 50-year leases, introduced after the Second World War to encourage trade and to prevent the Japanese buying Siberian timber. Federal losses in the Tongass have sometimes exceeded US $50 million a year. According to forest policy analyst Randel O'Toole, some US $330 million was lost over 15 years.

Forestry in the Tongass is uneconomic. For the last half century it has been heavily subsidized by the US government for a range of political reasons, including appeasing local interest groups and keeping the Japanese out of Siberia (a ploy which has now failed). Logging on Native American corporation land has sometimes been profitable in the short term, but only by logging unsustainably, thus effectively liquidating all assets, and because of additional government incentives such as the Net Operating Loss.

The Tongass was exempted, for many years, from most protection laws relating to the northern forest. In 1980, the Alaska Lands Act substantially increased the economic incentives for logging, including stipulations that the Forest Service provide an unprecedented 4.5 billion board feet a decade to the

Tongass timber industry. To aid this, Congress agreed to provide at least US $40 million direct subsidy a year for road building and infrastructure. Although the Lands Act designated substantial wilderness areas in the Tongass, most were in the higher, unforested regions. Some 91 per cent of timber was available for harvesting[146]. The reason for this succession of tax breaks, sweetheart deals and direct support is the extremely poor economics of forestry in southeastern Alaska. According to Wilderness Society calculations, some 93 cents are lost from every tax dollar put into the area, and in bad years this has risen to a 99 cent loss. A 1988 report from the US General Accounting Office found that during the first six years of Section 705 the Forest Service had spent US $131 million on preparing timber sales that no-one had bid for[147]. It is said that Douglas-fir has been sold to the Japanese for 1 per cent of the costs from the Pacific Northwest[148].

Despite all the support, employment in forestry has continued to decline, due to falling demands from some markets and competition from other areas. This has resulted in a rising spiral of federal losses from the Tongass. The Wilderness Society has calculated that in 1986 (a period of peak logging activity) federal losses translated to an annual cost of US $36,000 for every job in the logging and milling industries[149].

During the late 1980s, it became clear that areas of the Tongass were being cleared at an unsustainable rate, and irreplaceable old-growth rainforest destroyed. The Tongass also became a political embarrassment, with its special deals supporting Japan, the main USA trade rival. When US senators went to Brazil to urge President Sarney to reduce burning of the Amazon rainforest he countered by asking them what they were doing about the Tongass rainforest. In 1990, the Tongass Timber Reform Act introduced additional protected areas and some restraints on federal funding. It repealed the automatic US $40 million annual grant from Congress and eliminated the 4.5 billion board feet/decade target for the Tongass. About 296,080 acres of additional wilderness area was created and an additional 722,482 acres of forest put off-limit for logging. It also introduced a number of management changes, such as the leaving of 100-foot forest edges along fish streams. However, the 1991 target for logging was apparently the highest for a decade, and the buffer zones are thought to be inadequate by ecologists. For example, changes in microclimate measured back 787 feet in one study, meaning that the 100-foot buffer is not large enough[150]. An inter-agency task force assigned to assess risks to species found that over the next 150 years, species in several ecological provinces in the Tongass will be at moderate to high risk of extinction due to logging.

Other important Alaskan forests
Other important areas include the Kenai Peninsula and Prince William Sound, which now risk heavy logging from the Native American corporations, which have generally logged out their leases in the Tongass. There are some

indications that the corporations would be prepared to sell some of these areas for reserves if the money were available[151].

Small-scale logging, replanting and management also take place within the interior. Some of this involves management of state lands, where there is a policy of increasing management:

> Alaska's state commercial forest land base has declined due to allocation of forest land to uses that preclude timber harvest. . . Our goal is to make as much forest land as possible available for resource production. Present planning processes include the Susitna Valley, Tanana Valley State Forest, Kenai Peninsula and Takataga[152].

This means that there is a state aim to optimize production outside state parks and other protected areas. However, the finances of this are poor in the interior, with fairly low-value timber and poor access. Extraction is increasingly managed by use of winter roads, ie temporary roads laid on a snow base. Although it is claimed that this has no impact on vegetation, lack of maintenance can lead to erosion and slews[153]. A few exotics are also planted on a small scale, such as Siberian larch[154].

Logging has been fairly small-scale in the interior in recent years, but two large TNCs, ITT Rayonier and Minasha from Japan, are now looking at the potential of chipping or making chopsticks. The development is unlikely to proceed without substantial funding from the state for road building, but this is possible under the present administration.

CASE STUDY: SURINAME

Suriname offers an example of remaining tropical forest areas at risk. The population is small, and the country is densely forested, with 80 per cent of the total land area made up of tropical moist forest comprising one of the largest remaining stretches of natural forest left in the world. Current deforestation rate is around 0.1 per cent per year, one of the lowest in the South.

However, in 1993 a series of financial crises, and the threat of complete economic collapse, persuaded the government to approach timber companies in Asia for investment in logging industries. As a result, in early 1995 five major Asian companies from Malaysia, Indonesia and China, had put bids in to log around 3.6 million ha, or a quarter of Suriname's total land area, mainly in the middle of the country. The five companies include Berjaya of Malaysia, NV Mitra Usaha Sejati Abadi (MUSA) and Suri Atlantic of Indonesia, and two smaller Chinese state companies, including a consortium called Suriname Natural Resources Industries. Canada's Gordon Capital Corporation has also reportedly offered US$25 million to 45 million for a 51 per cent share of the state's own timber company, Bruynzeel, which already controls 470,000 ha[155].

There are questions about the quality of forestry practiced by several of these companies. In July 1994, Tony Yeong, the Managing Director of Berjaya Group Limited in the Solomon Islands, was expelled for allegedly attempting to bribe the Solomon's Minister of Commerce, Employment and Trade, although he claims that he was simply offering a gift and has resigned from the company. MUSA, which is already operating in Suriname and is apparently a consortium of about 18 Indonesian companies, has repeatedly been criticized for its environmental practices within the country[156].

Logging will not be easy. Much of the area is controlled by a mixture of Amerindians and 'Maroon' tribes, the latter being descendants of escaped slaves who have led a basically African tribal existence in the area for several hundred years. Both groups are armed and recently emerged from a long war with the government over land rights. They have apparently not been consulted about the proposed deal, and observers believe that the companies will have to fight their way in to extract timber[157].

If the concessions go ahead, they will totally alter the timber trade and ecology of Suriname. The three largest concessions would increase the country's total roundwood production by 15 to 20 times and increase export volume 300 to 350 times. The proposed concessions total over 143 per cent of the area of all Suriname's current domestically-owned concessions, creating the need for extra monitoring and regulatory capacity that currently does not exist. They will require extensive new road building, in areas where a hilly terrain means that there is a particular risk of soil erosion. The companies are intending to increase the rate of cut to include export of rarer species, in addition to the twenty or so that have already been identified[158].

Analysis by the Washington-based World Resources Institute (WRI) suggests that such short term measures will not even provide the government with an efficient source of revenue. The investors are currently offering about US$500 million, which is almost equal to Suriname's gross national product. However, most of Suriname's income would flow from corporate tax, which the companies could easily evade (and which foreign logging companies frequently do avoid throughout the tropics). Suriname would be likely to raise only a quarter of the potential revenues at most and the WRI analysis suggests that it could end up a net loser once costs of insuring compliances were factored in[159].

By July 1995, decisions about the future of the concessions remained on a knife edge. Opposition to the plans had grown within the country, and within the parliament, with opposing MPs delaying a vote. The MUSA bid had suffered badly because of criticism of its existing concessions, which have flouted local laws. The best advanced bid, from Malaysia, looks like being reduced in size and extent even if it goes ahead. Meanwhile, donor agencies have also been active in the debate. An agreed FAO project, due to start in August 1995, aims to strengthen the forestry sector within the country. A proposed US$5 million grant from the Netherlands will be aimed specifically at conservation measures, and additional money is potentially available from the European Commission, although the terms of reference have been criticised by NGOs. In a potentially significant development, Suriname ratified the Convention on Biological Diversity in July 1995, thus opening the way for further funding and strengthening the 160 arguments of those opposed to the logging[160]. Indigenous and Maroon leaders met in August to declare their opposition to the scheme. Meanwhile, similar plans are being advanced by Asian companies interested in logging in neighbouring Guyana, where short and long term gains from forest resources are being debated[161].

CASE STUDY: CAMEROON

Cameroon has been a significant exporter of tropical hardwoods for several decades. In the last few years, market pressure caused by depletion of West African forests, coupled with a government commitment to increase logging, have together resulted in a rapid increase in the rate of cut.

At the beginning of the 1990s Cameroon was rated the seventh largest tropical timber exporter in the world, and third largest in Africa, after Côte d'Ivoire and Liberia[162]. However, since then there has been a steep rise in logging, and the country has almost certainly overtaken Liberia in level of exports[163], with timber cut increasing by 100 per cent over a couple of years[164]. These changes reflect the Cameroon government's intention to be the largest timber exporter in Africa by the year 2000, an aim that has been described as 'completely unrealistic' in terms of a sustainable cut by a study for the International Tropical Timber Organisation[165]. However, rising prices on the South East Asian timber market have persuaded companies to increase the rate of extraction. The increase has also been at least in part encouraged by the National Forest Action Plan produced under the auspices of the Tropical Forestry Action Plan, which was criticised in a report to WWF for encouraging unsustainable rates of logging[166].

Domination by foreign companies

Like many West African countries, the timber trade in Cameroon continues to be dominated by foreign companies, including both TNCs specifically involved in timber, and those for which wood products are just one part of a larger operation. This in turn means that the majority of exports remain as raw logs, destined for conversion in Europe.

Over 60 foreign companies are currently operating in Cameroon[167]. French companies are by far the most prominent, with at least 18 operating between 1985 and 1990[168], the five largest being Becob, Rougier, Rivaud, Thanry and SCAC. Some important European companies operating in Cameroon include:

❏ **Decolvenaere** (Belgium) through SFIL (Société Forestière et Industrielle de la Lokoundje) and SOTREF (Société Tropicale d'Exploitation Forestière du Cameroun)
❏ **Becob and Isoroy** (France) through SOFIBEL (Société Forestière et Industrielle de Belabo)
❏ **Pasquet** (France) through PALLISCO
❏ **Rivaud** (France) through EFC (Enterprise Forestière Camerounaise) and La Forestière de Campo
❏ **Rougier** (France) through SFID (Société Forestière et Industrielle de la Doumé)
❏ **SCAC** (France) through SIBAF (Société Industrielle de Bois en Afrique)
❏ **Thanry** (France) through SEBC (Société d'Exploitation des Bois de Cameroun)
❏ **Danzer** (Germany) through GRUMCAN (Grumes de Cameroun)
❏ **Feldmeyer** (Germany) through CIFOA (Compagnie Industrielle et Forestière de l'Ouest Africain)
❏ **Alpi** (Italy) through ALPICAM
❏ **Itallegno** (Italy) through ECAM (Compagnie d'Exploitation Industrielle des Bois du Cameroun)
❏ **Reysir** (Italy) through COCAM (Les Contraplaques du Cameroun[169].

Specialists who have studied forestry operations in Cameroon conclude that there is, at present, effectively no sustainable forest management taking place. Major problems include poor road building leading to soil erosion, construction of unnecessary skid trails, high wastage in forestry operations and unfeasibly short cutting cycles[170]. Foreign companies have frequently been criticized both for their forestry practices and the legality of their operations. For example, a report in *The Times* in 1990 stated 'it is an open secret that a large proportion of the [foreign-owned] companies are not paying their government taxes'[171].

Resentment against foreign timber companies is increasing within Cameroon. Operations of SFID, the largest timber company in Cameroon which is affiliated to the French company Rougier, prompted villagers in Mbang to write a protest letter which stated that 'a company like SFID, having important financial means, seems to pay no attention at all to the misery of the people and even neglects its obligations towards the population'. Complaints include the claim that the company burns timber waste, rather than giving it to villagers for construction or fuel, despite repeated requests, and that legal requirements about logging areas and minimum tree sizes are ignored. A particular area of concern is about the fate of the moabi tree (*Baillonella toxisperma*), which is used by local people as a source of fruit and to manufacture karite oil from the seeds; the latter provides a cash crop for some villagers and is highly prized. The trees are now being logged. A Dutch study reports an SFID manager as saying that 'if we see a moabi, we'll log it, no matter if the trees are felled in the neighbourhood of settlements or below the minimum diameter of 1.20 metres'[172].

Deforestation appears to be increasing at an enormous rate, and has risen by 400 per cent over the last few years. High prices for timber on the international market, particularly in parts of Asia, has encouraged timber companies to increase the rate of cut. Recently, a Lebanese company began logging on a 100,000 ha site on the edge of the Dja nature reserve[173]. Unless changes in management policy can be agreed within the next few years, native forests outside reserves are unlikely to survive long into the twenty-first century.

CONCLUSIONS

Far from being negligible, the impact of the timber trade on natural forests is important and in many cases decisive. In temperate and boreal regions, it is by far the most important single cause of forest degradation and loss. In tropical regions, it is a major cause of loss of many of the world's most important remaining forests in Southeast Asia, the Amazon and West Africa. Logging of natural forests does not take place by accident. It occurs because natural forests are undervalued and underprotected. Addressing the issue of natural-forest loss from the timber trade is perhaps the main concern, with respect to forest conservation in many parts of the world.

5
Intensification of Management in Secondary Forests

With the exception of a few reserve areas, nearly all the accessible forest in the world will soon be under management. Systems of management vary markedly, and range from a virtual hands-off policy of minimal maintenance for biological reserves and national parks, to intensive, short-term plantations of non-native species. Worldwide, the tendency is to intensify management of forests. This has serious environmental and social implications.

The future of forest policy is often presented as a switch from logging in natural forests to extracting wood from managed forests. This is an oversimplification. Few, if any, of the world's forests are totally unmanaged; the exceptions may be the tiny minority that are all but inaccessible to humans, such as a few mountain plateaus in South America, or long-uninhabited islands. Even vast 'natural' forests, like the Amazon and the boreal forests of the far north, have been subtly (and sometimes not so subtly) altered by factors such as the actions of indigenous people in tropical forests, local herding of reindeer in Lapland and Siberia, and use of fire by aboriginal groups in Australia and New Zealand. Given this context, it would be more accurate to say that what we are witnessing is a change from logging of near-natural forest to an *intensification* of management of secondary forest and conversion of semi-natural forests to plantations.

Forest management ranges across a broad spectrum. At the one extreme, it can involve taking a policy decision to leave an area of forest strictly alone, for example to provide protection of watersheds or nature conservation. On the other hand, it can mean intensive, three- to five-year rotations of closely packed non-native monocultures. A range of examples given in the following two boxes.

SYSTEMS OF FOREST MANAGEMENT

Forest management systems can vary from a decision to leave an area of forest completely alone to development of an intensively managed plantation. Important distinctions include choice of species and harvesting methods. All forestry methods include options for a greater or lesser degree of weed control, acceptance of natural regeneration, soil preparation, etc. In the examples below (which are not exhaustive) the primary distinction is made with reference to harvesting method. Harvesting is usually the most severe management impact and is thus a logical choice for using as the criterion for classification. However, management systems vary in more than harvesting method. The list below shows different strategies along with some of the major options facing forest managers in each one.

No removal
1   Land set aside for nature conservation, watershed protection, soil protection, etc

Some removal (selective or clearcut)
2   Land primarily for non-timber uses, some removal and planting necessary for enhancement of habitat:
  *   hunting, game-bird rearing;
  *   recreation;
  *   conservation, etc.
3   Removal of whole trees, or parts of trees, occasional enhancement planting:
  *   coppice;
  *   coppice with standards;
  *   pollards;
  *   removal of twigs and branches for firewood;
  *   selection systems.
4   Removal of trees, natural regeneration:
  *   through abandoning the site for some time;
  *   with added soil treatment by scarification or prescribed burning;
  *   with artificial fertilizer and pesticide inputs.
5   Removal of trees, with enhancement planting:
  *   with or without soil treatment;
  *   native or non-native trees*;
  *   natural or genetically engineered trees;
  *   different degrees and methods of weed control;
  *   with or without additional natural regeneration.
6   Removal of almost all trees, with some trees retained in, eg, fire refugia or isolated stands:
  *   options as in 5 above.
7.   Removal of trees with replanting:
  *   with or without soil treatment;
  *   native or non-native trees;
  *   natural or genetically engineered trees;
  *   different degrees and methods of weed control;
  *   different rotation times for trees.

* Non-native can include both foreign species and species found in the country but not in the particular area, including sometimes non-local provenances of local species.

REPLACEMENT STRATEGIES FOR MANAGED FORESTS

Following extraction methods (see Box 5A), the most significant impacts of forest management come from the method of restocking trees following clearance. Again, a number of options are available, of varying degrees of intensification.

Natural regeneration

The simplest and perhaps the commonest option of all is natural regeneration, ie leaving forest to regrow from existing saplings and seeds. Sometimes regeneration is aided by a number of soil treatment methods after felling, including *prescribed burning* or *scarification*. Burning helps mimic the effects of fire, aids some seed germination, releases nutrients in forestry waste and helps control certain pests and diseases, including fungal attack. It is the traditional soil treatment method in many temperate countries, but is now increasingly being replaced by scarification, the mechanical breaking of the soil surface to aid germination and tree growth. However, burning is not invariably the most benign method, particularly if forests are susceptible to invasion by weeds and alien plant species. In the sub-boreal spruce zone in Canada, burning was found to reduce the abundance of a number of original plant species on submesic and mesic sites, in part due to a rapid influx of fireweed (rosebay willowherb – *Epilobium angustifolium*).

   Natural regeneration is the method adopted, often through default, over wide areas of the tropics, following selective logging, unless the aim is to convert natural forests to plantations. In practice, land is often cleared for other purposes before trees can regrow. However, if the forest is allowed to regenerate, and if sufficient areas of standing forest remain to provide a reservoir of biodiversity, natural regeneration can in some areas result in an approximation to a natural-forest system. Examples of areas where it is practised today include Sabah in Malaysia, parts of British Columbia in Canada, and areas of Finland.

Enhancement planting

In some areas, particularly within temperate and boreal countries, natural regeneration is boosted by selective planting, or sometimes by seeding. Such a system will generally use native species although there is an increasing tendency to include alien provenances, ie native tree species but individuals that have been grown in different regions, or even different countries. Much contemporary planting of birch in Scotland, for example, is from seed produced in the Netherlands.

Replanting

Replanting, with either native or non-native species, is becoming the norm in industrialized countries, and is spreading rapidly in less developed areas as well. Replanting can vary between establishing mixed stands of native species in roughly the same proportions as would be found in a native forest, to planting monocultures of exotic species. In forestry, as opposed to recreational or conservation management, the tendency in many areas is towards the latter situation, ie a *plantation*. Even if native species are planted, the result is usually largely a monoculture with a considerably reduced understorey. Native species are unlikely to be of local provenance.

Afforestation

Planting on areas that are naturally, or have for a long time been, unforested is also an increasing phenomenon, especially in Europe and Australasia. Plantation establishment of this type tends to involve considerable site preparation, including often quite deep ploughing, and subsequent management. Planting is predominantly with non-native species.

Planting with genetically-engineered trees

Increasingly, forestry companies are becoming interested in the application of contemporary research in genetic engineering, and particularly tissue culture and cloning, to plantation establishment.

Source: This box draws on E H Hamilton and H K Yearsley (1988) *Vegetation Development after Clearcutting and Site Preparation in the Sub-Boreal Spruce Zone* ERDA, Canada and British Columbia, FRDA Report 018

CHANGING FOREST MANAGEMENT SYSTEMS

Forest management systems are driven by the needs of the dominant industrial interests at a particular time. For example, many traditional European coppice systems were developed to provide fuelwood, both for domestic purposes and for industries such as iron smelting in the period before widespread use of coal. Whole silvicultural systems were developed to produce oak (usually *Quercus robur*) of the right size and shape for shipbuilding[1].

Today, an increasing proportion of the forest estate is aimed not at producing *timber* as such, but at the rapid and continuous production of *fibre* for various forms of pulp and wood chips. This change has had a profound effect on forest management strategies, generally to the detriment of non-timber uses. For example, use of monocultures eliminates fruit trees, and often also sources of traditional medicines, herbs and fodder.

TRADITIONAL MANAGEMENT SYSTEMS

In many cultures, forest management has taken place by default; an area of forest is cleared and then left to regrow. When the human population is low, and forest resources relatively abundant, such a system works quite well and it is currently the basis of slash-and-burn agriculture throughout the developing world. However, it quickly breaks down when there is greater demand for land or timber, or when population pressure results in shortening the rotation to beyond the forest's regenerative capacity. The latter is now happening in many areas of the South. In northern Thailand, for example, rotations have been substantially shortened because of population pressure, both from the enlarging local population and as a result of migration from other areas, and the forest no longer has time to recover between fellings. Knowledge of replanting and management varies widely in traditional societies, from the existence of very sophisticated arboricultural systems to lack of even simple tree management skills.

If silviculture is practised at all, forest managers have various options open to them, depending on their particular requirements, the timber species and the conditions within the forest. Although much early timber utilization involved wholesale clear-felling, often of natural forests, selective removal systems were fully developed hundreds if not thousands of years ago in Europe. There, early problems of resource shortages and population pressure forced foresters to develop more integrated systems of management. Some common options, used particularly in temperate countries, are outlined below.

Coppice is probably the commonest selective system. It involves cutting the trunk at ground level, resulting in regrowth of multiple shoots from the base, which can be cut again every few years. Traditionally coppiced species in Europe include hazel (*Coryllus avellana*), oak (*Quercus* spp), alder (*Alnus glutinosa*), ash (*Fraximus excelsior*) and willow (*Salix* spp). The coppice cycle is usually between 3 and 20 years, so that productivity is more regular than when full-size trees are felled. In traditional systems, such as those described in Thomas Hardy's novel *The Woodlanders*, an area of forest would be managed so that some cutting could be carried out each year. Some species can be successfully coppiced many times; for example oaks (*Quercus robur*) established as coppice in the Middle Ages are still living in parts of Europe. Coppice was traditionally used for fuelwood and charcoal, poles for building, tool handles and fencing. Today, it is being investigated again as a potential source of biomass for fuel and/or fibre.

A common variation is **coppice with standards**, whereby most trees are coppiced but some are established to grow to maturity, the latter being selectively removed for their timber. Many different combinations can be used, a popular one being oak and hornbeam (*Carpinus betulus*). Coppice with standards is a good method of obtaining high quality timber from the trees which reach maturity, because shade from the coppice trees helps suppress side branches, thus creating straight, strong trunks.

Pollard management is similar to coppicing, although more limited in its application. It involves cutting young branches at a height of 2 to 4 m, creating a bushy regrowth. Pollarding is often used with willow, where branches can be cut every year for fencing, basket-making, etc. It is often used on water meadows or where browsing by livestock precludes coppicing.

MODERN MANAGEMENT SYSTEMS

Traditional management systems were overturned by a series of developments in Europe at the end of the nineteenth century. Faced with serious deforestation in parts of central Europe and Scandinavia, a new and more intensive model of forestry was developed over a number of decades. The new model is usually referred to as the *sustained yield model*, or simply as the *Scandinavian model*, although it was actually first developed in Germany. Its current association with Scandinavia is due to the fact that it has been applied and refined during the past 40 years particularly in the Nordic countries.

The sustained yield model is characterized by a primary concern with timber production and therefore with the maintenance and increase of on-site woody biomass. It is based on careful assessments of the potential for continuous productivity from a given area of forest over time. Harvest must be less than or equal to timber growth, and therefore a critical factor in operation of the sustained yield model is a knowledge of the rate at which trees of a particular species grow on a given site. This is usually measured in terms of cubic metres of tree growth per acre or hectare per year and is reflected in management plans by controls such as the annual allowable cut.

In theory, a sustained yield model can, and indeed sometimes is, used in a natural forest, where small amounts of timber are extracted, and the forest is allowed to replenish itself through natural regeneration. Sustained yield models are also in principle easily adaptable to multiple-purpose management, where timber production takes place alongside other uses such as hunting, food gathering, recreation and nature conservation. However, in practice, the emphasis has been placed, to an ever greater extent, on the production of timber in preference to other goods and services.

For the past few decades, the principles of sustained yield forestry have held a virtual monopoly in attitudes towards forest management. Once the model is established, the main focus for improvement is towards *increasing* annual growth, and management has been directed to that particular, and fairly narrow, end. Success has been measured in terms of the *sustained yield* of timber or biomass. Over the last 50 years, this has come to be achieved by means of a number of management practices that have assumed a position of such prominence that it has been almost impossible to challenge them within conventional forestry circles. They include:

❑ replacement of mixed forests with monocultures;
❑ use of exotic species which grow particularly quickly on a given site;
❑ use of native species but with seed brought from other countries or regions;
❑ use of nursery grown seedlings;
❑ breeding programmes of commercial tree species aimed at increasing productivity and improving form;
❑ harvesting by clearcutting;
❑ elimination of competing tree species, known as 'weed species';
❑ draining of wetland sites and mires;
❑ diminished ecological processes, including greatly reduced nutrient cycles;
❑ simplification of complex ecological systems;
❑ suppression of fires;
❑ a drastic reduction in rotation length (the age at which trees are felled and replanted);
❑ use of artificial fertilizers to increase yield;

Table 5.1 Plantations in the tropics, by region

Region	Plantation area (ha in 1990)	Annual increase 1981-90 (ha)
West Sahelian Africa	251,000	21,000
East Sahelian Africa	762,000	32,000
West Africa	445,000	14,000
Central Africa	175,000	11,000
Tropical Southern Africa	1,057,000	47,000
Insular Africa	310,000	4,000
South Asia	19,758,000	1,480,000
Continental Southeast Asia	3,197,000	140,000
Insular Southeast Asia	9,156,000	482,000
Central America and Mexico	273,000	17,000
Caribbean subregion	442,000	23,000
Tropical South America	7,922,000	333,000
Tropical Oceania	43,000	2,000

Note: Some of these may be underestimates due to lack of data from some countries. In particular, the Tropical Oceania figures are derived solely from Papua New Guinea.

Source: N Dudley and S Stolton (1995) *Pulp Fact: The Environmental Impacts of the Pulp and Paper Industry*, WWF International, Gland, Switzerland; figures derived from A L Hammond, (1994) *World Resources 1994–95*, World Resources Institute with UNEP and UNDP, Oxford University Press, Oxford and New York

❏   use of pesticides, including especially herbicides to suppress weeds and insecticides sometimes applied aerially;
❏   increased use of machinery for logging, harvesting, etc; and
❏   use of scarification and deep ploughing methods in soil preparation before planting.

The popularity of the sustained yield model

At its extreme form, intensification means the replacement of a natural forest with a plantation. These changes have had dramatic implications for forest ecology, as outlined below. However, until recently the sustained yield model has largely been accepted as equivalent to environmentally sound management, particularly in temperate and boreal regions where the majority of such management has taken place. Although a successful publicity campaign by timber companies has helped gloss over inconsistencies within management, their task has probably been made easier by an accident of semantics. Following publication of the Brundtland report *Our Common Future*[2], *sustained yield* in a forestry perspective became virtually synonymous with a wider concept of *sustainable development* or simply *sustainability* (a word that has gained a high profile in the last few years but still fails to occur in most dictionaries). Sustained yield thus fitted neatly into a popular new concept, and until

Table 5.2 Environmental problems caused by tree plantations

Problem	Notes
Loss of habitat	In many places, plantations have been established in the place of existing natural forest and natural or ancient heathland, etc. Some of the habitats replaced are themselves now scarce and requiring conservation
Reduced biodiversity: trees	Plantations tend to use exotic tree species, or selected strains of native trees (sometimes through use of tissue culture to develop identical trees), thus reducing genetic variability and hence adaptability of trees
Reduced biodiversity: other wildlife	Replacing native or managed forests with plantations inevitably leads to a substantial reduction in biodiversity, especially with respect to lower plants and invertebrates
Introduction of exotic tree species	Exotics sometimes compete with native species, and have occasionally hybridized with native species, leading to loss of local provenances
Soil erosion	Deep ploughing during planting, especially in upland areas, can lead to excessive soil erosion, as can the impact of clear-felling
Acidification	Establishment of conifer plantations on base-poor soils has led to increased acidification of both soil and freshwaters, due mainly to the role of conifers in scavenging air pollutants, which are later washed down trunks, although conifers may have an additional acidifying role of their own
Water table changes	Planting some species, eg eucalyptus, can lead to serious lowering of the water table in drought-prone areas
Water quality changes	Drainage, ploughing and clear-felling can all lead to an increase in water turbidity, which can damage the breeding success of salmonid fish, disturb other aquatic life, etc
Changes to the fire ecology	Plantations can either increase fires, through poor management, or artificially suppress fires. Both can have serious ecological consequences
Increased pest and disease attack	Monocultures are particularly prone to attack by pests and diseases. Introduction of exotic trees has sometimes resulted in parallel introduction of serious pests
Agrochemical use	Pests and problems of decreasing fertility lead to greater use of pesticides and soluble fertilisers; both of these have a range of environmental problems including water pollution, damage to wildlife etc

Source: This table first appeared, in slightly altered form, in N Dudley (1992) *Forests in Trouble: A Review of the Status of Temperate Forests Worldwide* WWF International, Gland, Switzerland; additional sources include J Sawyer (1993) *Plantations in the Tropics*, IUCN, UNEP, WWF, Switzerland; G Rosoman (1994) *The Plantation Effect* Greenpeace New Zealand, Auckland

recently appealed both to industrial interests and to sections of the environmental community.

Indeed, foresters in many industrialized countries still look towards the model as an ideal. Its promotion by Scandinavian-based international consultancies, such as Jakko Poyry and Swedforest, has spread the message into tropical countries, and more recently into central and eastern Europe as well. A report of a study tour of Scandinavia by Canadian foresters praised the clearcutting methods, which they noted appeared to have been an issue ten years earlier but 'now seems to be generally accepted by the public'[3]. The authors recommended that the British Columbian forestry industry should focus 'more on environmentally sound management practices than on alienating more of the working forest as a means of preserving the environment'. 'Alienating' here means putting into reserves or national parks.

The intensive management model is being adopted in the tropics as well. There, managed forests are frequently plantations, often using introduced species such as acacia and eucalyptus. While remaining a tiny proportion of the managed area of the temperate and boreal regions, tropical plantations are increasing in area very rapidly. The faster growth rate of trees in tropical countries means that, in trade terms, countries such as Malaysia and Indonesia could easily be providing opposition to temperate countries in a relatively short space of time. Some tropical plantations are summarized in Table 5.1.

ENVIRONMENTAL AND SOCIAL ISSUES RELATING TO MANAGED FORESTS

Despite its long development, sustained yield is still basically just a measure of how efficiently timber can be produced from a given site on a continual basis, ie crop yield. It does not take into account any other forest benefits such as biodiversity, environmental services and wider social, economic or cultural values. As a result, by concentrating on timber or biomass alone, many important goods and services of the forest have been ignored or greatly undervalued, such as the quality of fisheries, watersheds and wildlife, and climatic stability.

Any form of land management changes natural systems. Land set aside primarily, or even partially, for the production of timber will not have the same ecology, nor will it meet the same range of human needs, as a fully natural forest. (It may, of course, meet some human needs better than a completely untamed area of forest.) However, modern forestry management, as dictated by the perceived needs of the timber trade, has created such unnatural systems that forest areas are actually sometimes net drains on the environment, rather than contributors to environmental stability.

Modern management systems, with their requirements for maximizing production at the expense of timber quality, have concentrated for the most part on clearcutting. Indeed, complicated rationales for clearcutting have been developed, such as its being necessary for regeneration, or the most 'natural' form of disturbance on a given site. (There may be cases when this is true, although not in the manner in which clearcuts are practised in most places today; this is discussed in the section on management options in Chapter 7.) Once cutting has taken place, there are a number of strategies for ensuring that forest regrows on the site. Table 5.2 summarizes some of the major issues associated with intensive forest management and the establishment of plantations. These questions are expanded below.

Monoculture establishment and biogeographical effects

Throughout the world, there is a tendency for natural, mixed forests to be replaced by monocultures. In addition, native tree species are being replaced by a narrow range of high yielding varieties, consisting mainly of conifers and specialized broadleaved trees such as *Eucalyptus* and *Acacia*. Over increasingly large areas of the planet, 'forests' are actually plantations of non-native species.

In temperate regions, a few conifer species dominate plantation forestry. These include silver fir (*Abies alba*) from central Europe; Douglas-fir and western hemlock from North America; Norway spruce (*Picea abies*) from northern Europe; Sitka spruce from northwest North America; radiata pine (*Pinus radiata*) from a small area in California; European larch (*Larix decidua*) from central Europe; and Japanese larch (*L kaempferi*) from Japan. Other popular plantation species include various *Eucalyptus* from Australia and southern beech (*Nothofagus*) from New Zealand and Chile. In parts of Europe and Asia various poplar (*Populus*) species are also widely planted. In the tropics, although plantations are nowhere near as well developed, pulp plantations are concentrating on such species as oil palm (*Elaeis guineensis*), acacia and eucalyptus, along with conifers.

The proportion of exotic trees in the landscape depends on government policy and the ambitions of the timber trade. Generally, replacement has proceeded fastest in deforested temperate countries, although this trend is now changing. Originally often government driven, private companies are now frequently reaping the benefits of what were previously state-driven afforestation schemes. The majority of forest cover in a number of European countries is now of non-native species. In Denmark, for example, 66 per cent of the forest cover is conifer plantation and forest area has increased from 2 to 12 per cent since 1800[4]. Scotland has 15 per cent forest cover, 93 per cent of which is commercial conifer plantations[5], mostly of non-native species. In Spain, *Pinus insignia*, *Pinus radiata* and *Eucalyptus globulus*, planted since 1950, now cover over 5 million ha, while hardwoods occupy only 1.9 million ha[6]. Some tropical countries are now rapidly catching up. In Chile there are 9 million ha of natural forest and 1.3 million ha of plantations, mainly of *Pinus radiata*. The area of the latter is expanding rapidly[7] following investment by both Chilean and foreign timber companies. In the European Union, a Directive now virtually ensures that non-local oak is planted by insisting on certain seed being used to ensure straight trunks; for example in the UK most oaks planted in commercial plantations will come from acorns collected in Hungary and Poland[8].

Impacts on biodiversity and ecology

Intensive forest management also results in the disappearance of wildlife within the forest system. The degree to which native plant and animal species can utilize managed forest habitats depends to a large extent on how similar they are to natural forests. While some traditional systems have a high associated biodiversity, many intensively managed forests and plantations do not. Almost all managed systems have a *simplified* biodiversity, ie although numbers of species may be high they are likely to miss out some important components of a natural forest, and probably introduce some potentially damaging non-native or non-local species.

As rotation times are shortened, climax forests (ie the oldest trees) and associated species disappear from the landscape. This means that species reliant on old-growth characteristics can no longer live in managed forests. Although a minority of species get all the publicity, like the spotted owl in North America and the white-backed woodpecker in Europe, the real victims of old-growth loss are likely to be smaller and more obscure species which do not capture the eyes of forest ecologists or the imagination of the public. The extent of biodiversity loss has often been obscured by concentrating research on larger animals, such as birds, a proportion of which can adapt to even quite regimented plantations. Arguments in support of clearcutting have been that overall numbers of species can *increase* following felling. However, a simple head count masks the fact that rarer and more ecologically sensitive species often disappear following clearcutting, to be replaced by invasive pioneer species, usually including introduced species.

As a result of these changes, the development, composition and structure of the forest is simplified. There will generally be fewer tree species, and thus less associated wildlife. An altered balance between coniferous and deciduous trees is likely, and the proportion of standing dead trees (snags) and down logs will decline dramatically, along with associated flora and fauna. Recent research from the boreal regions of Europe suggests that up to 25 per cent of biodiversity in some forests may be associated with dead wood which is likely to decline or disappear in managed forests[9]. Studies in the UK have found that a greater number of insect species are likely to be found on native rather than introduced trees[10]. The Scottish crossbill (*Loxia scotia*) has evolved particularly to live in single-species conifer forests of Scots pine (*Pinus sylvatica*) which are the natural habitat of the region[11], and its habitat is reduced by the introduction of other conifer species in plantations.

Fire ecology will also be drastically altered. In many forests, this has led to greater frequency and severity of fires, resulting in environmental damage, wildlife loss, economic costs and social distress[12]. However, in other areas, especially in some northern temperate forests, excessive fire suppression is also causing problems by gradually altering the forest mosaic and leading to build-up of pest species[13]. Sometimes, frequent ground fires have been replaced by stand replacement fires due to suppression activities[14]. A substantial propor-tion of the most endangered species in Finland are those which need fire and ash, and have therefore been threatened by management policies that reduce incidence of fire[15].

One important side effect of introducing new species of trees is that it adds to the risks of weed invasion into natural areas, both by the trees them-selves and by herbaceous or fungal weed species introduced accidentally at the same time. This problem is particularly well known in Australia[16] and New Zealand but occurs in nearly all forests. An estimated 560 alien plant species are found in New Zealand and 240 are common throughout the country[17]. Most have not invaded unmodified forests but some have become important in early forest succession or in forest edges. Similar problems are experienced in central Europe from black locust (*Robinia pseudoacacia*), originally from North America, which has been extensively planted as a timber tree and later spread as a weed along roads, forest edges and corridors. In Africa, *Cecropia* trees from Asia are starting to invade natural areas.

Establishing forestry on previously unforested or deforested land can also be damaging, particularly if exotic species or insensitive planting methods are used. In the UK, previously uncommon species have been put further at risk in some areas by conifer afforestation, including the hen harrier (*Circus cyaneus*). In 1992, the European Community prosecuted the UK government for allowing conifer afforestation in the Flow Country of northeast Scotland without carrying out a full environmental impact assessment[18].

Soil and water effects

Forest management can affect hydrological cycles through both deliberate manipulation and accidental side effects. Intensification of forest manage-ment in many northern temperate forests has involved drainage of wet forests and natural peat areas. In Finland, for example, some 6 million ha of natural peatlands, 60 per cent of the total, have already been drained for forestry[19]. In the UK, large areas of globally rare blanket bog (*Sphagnum*) habitat have been destroyed through afforestation[20]. In these areas, hydrological systems are also upset by ploughing, scarification and brash burning, which can open up soil to erosion by wind and rain and increase gulley erosion on steep slopes. Sediment load in upland Britain was found to increase fifty-fold after drainage and took some years to settle down, to a level that remained at four times the original level[21].

In other places, afforestation has the opposite effect because trees take too much water, drying out soils. For example, it has been suggested that eucalyptus plantations in dry areas of India and southern Europe have been altering local hydrological systems[22]. Detailed research in Karnataka state suggests that in some situations eucalyptus takes up no more water than indigenous species, but in other cases water use is greater, posing possible threats to long-term eucalyptus cultivation. In parts of Australia, eucalyptus roots have penetrated into groundwater sources[23].

Regular timber removal also means a constant export of nutrients from the system, particularly in the case of whole-tree harvesting, including losses of nitrogen, phosphorus, potassium and calcium[24]. Changes in management may reduce nutrient availability. For example, input of nitrogen into Pacific Northwest conifer forests occurs through the actions of nitrogen-fixing organisms in early or late successional stages of forest development. Short cutting cycles and suppression of other vegetation in intensively managed plantations may have the net effect of reducing nitrogen input, although the full impacts have not yet been measured[25]. Conventional management practice assumes that, once nutrient deficiency starts to affect tree growth, this will be made up by use of soluble fertilizers, although the long-term efficacy of this is yet to be proven. It has been estimated that the loss of phosphorus from three pine rotations on an infertile soil is the equivalent of 20,000 years of natural loss and 1000 years' natural loss on recently fertile soil[26].

Tree health

Changing forestry methods have led to increased pest attacks. Uniformity of species in monocultures or other plantations makes it easy for pests to spread quickly. Clear-felling that results in substantial windthrow around the edges of clearing could provide ideal conditions for spread of bark beetles in parts of the Pacific Northwest region of North America[27]. Sometimes introduced pests also become important, such as the North American longhorn beetle which now thrives in Europe; Dutch elm disease (caused by the fungus *Ceratostomella ulmi*) which has wiped out most of Europe's elm trees and has been traced back to the Himalayas[28] and damaging nematodes which may spread through imported timber[29]. Control methods are admitted to be partial and inadequate. The most widely distributed and destructive defoliator on coniferous trees in western North America, the western spruce budworm *Choristoneura occidentalis*, has caused serious economic losses over large areas. Oversimplification of habitat through plantation forestry reduces the habitat for the natural enemies of pests, for example woodpeckers and insectivorous bats.

Pollution and impacts on air and water quality

Modern forestry makes heavy use of soluble fertilizers and a range of herbicides, insecticides and other pesticides. These all have effects both within the forest ecosystem and on surrounding natural or semi-natural habitats, including watercourses. Nitrate fertilizers cause serious problems of eutrophication in freshwater systems[30]. Herbicides can kill non-target plants and are sometimes directly toxic to animals including the natural enemies of pest species[31]. During pest outbreaks, insecticides are frequently applied through aerial application, with a high risk of drift. Many agrochemicals passed for use in forestry areas have known or suspected health effects on humans and wildlife[32]. For example, fenitrothion and other pesticides have been applied against spruce budworm on forests in eastern Canada for over 40 years. Canada's Future Forest Alliance estimates that spraying 1600 km^2 could result in the deaths of over 175,000 migratory songbirds, along with countless insects, spiders and aquatic creatures[33]. The herbicides glyphosate and hexazionone are used in the boreal zone of North America[34]; glyphosate is extremely toxic to broadleaved plants and can cause damage to wild plant communities. There is increasing evidence that pesticides can damage the health of forestry workers and others in the vicinity of spraying, especially from using persistent or toxic chemicals such as DDT, 2,4,5-t, 2,4-D[35]. In New Zealand, more than 30 brands of chemicals are used on plantations, with some areas being treated more than three times a year[36].

Conifer plantations can increase acidification in catchments through their ability to capture air pollutants such as sulphur and nitrogen oxides, which are deposited on trunks and needles. These are later washed to ground as acids by rain and snow, and affect soil chemistry and freshwater systems[37]. Measurements in the headwaters of the Severn and Wye rivers in the Plynlimon mountains in Wales found that watercourses became 14 times as acidic than they had been before afforestation[38]. Acidification can kill sensitive fish and invertebrates, and can affect other wildlife, such as the dipper (*Cinclus cinclus*), an upland water bird which is declining because acid waters do not support sufficient invertebrates for it to feed upon[39].

Social implications of forest management

Intensive forest management also has a number of important implications for people living in or near forests and for people who visit the area for sport or recreational purposes. A number of issues can be identified[40]. On the positive side, forestry can provide direct financial and resource value and/or employment, including a range of forestry operations and sales of timber products, although in general employment per given area of forest is declining in the North. Also important are indirect employment through forestry, including timber processing, pulp mills and local timber users, and the maintenance of a sufficiently large rural population to justify services such as shops, schools, medical facilities, garages, village amenities, public transport, etc.

However, forests also provide other forms of employment or financial value, including acting as grazing areas, sources of fuelwood, building materials, fodder, berries, mushrooms, game and herbal medicines. Many of

these uses decline under systems of intensive management. Tourism development, and its related spin-offs, is likely to be more successful in an area of
natural or near-natural forest than in intensively managed areas. Increasingly,
people perceive intensively managed forests as an obstacle or a problem rather
than a net benefit. Potential problems include:

❑ conflicts over land use arising from growth or decline of particular forest
   activities, including desires of local communities as compared with
   visitors;
❑ loss of land by indigenous peoples;
❑ disruption of communities through forest activities, including temporary
   immigration of labour, long-term immigration of outsiders with different
   background and outlook, social implications of increased tourism;
❑ disruption of neighbouring land uses through hydrological problems, soil
   erosion, etc;
❑ loss of land for grazing, hunting, fuelwood collection, etc;
❑ the forest as a barrier to access (eg to uplands, fishing, etc);
❑ the forest as a source of inflammable material;
❑ negative landscape value of the forest;
❑ aesthetic and practical problems arising from forest uses, including resentment towards forest management, changes in public access, etc;
❑ trespass, litter, damage and road congestion arising from more visitors to
   forests and woodlands;
❑ trees as barriers to agriculture, transport systems, etc;
❑ pollution from tipping and agrochemicals used in forestry;
❑ loss of aesthetic, cultural or spiritual values attached to natural forests; and
❑ loss of local distinctiveness in forests.

Many of these problems can be addressed, and indeed forests can be seen
as a positive advantage, if local people are given a stake in the forest and a role
in decision making and management. However, the economics and political
structure of large-scale plantation development means that such opportunities
are often missed in practice.

THE FORCES BEHIND INDUSTRIAL FORESTRY

Forest management has become big business. Traditional timber
companies and many other transnational corporations are becoming involved
in a range of forestry ventures. Most of the larger timber companies are
involved in intensive forest management, in setting up entirely new plantations
or in monoculture reforestation of logged over old-growth areas.

One particular phenomenon of the last few years is that a range of
companies with no previous interest in forestry are starting to become
involved, usually in search of feedstock for profitable pulp or chemical sub-

sidiaries, or as diversification from what they perceive may be declining markets. For example, the petroleum and chemical TNC Royal Dutch Shell has developed forest plantation interests in both tropical and temperate areas, including operations in the Congo, South Africa, Brazil, Chile, New Zealand and Thailand[41], with a particular emphasis on eucalyptus[42] and radiata pine, with a 20,000 ha plantation of the latter in Chile. The tobacco giant BAT has been involved in several plantation projects and retains an interest in a major plantation in Brazil run by Aracruz, although there are signs that BAT itself is withdrawing more firmly into tobacco again. Eucalyptus Paper Mills, which holds a 76 per cent stake in the Portuguese Companhia de Celulose do Caima, is controlled by the UK-based brick manufacturer Ibstock Johnson[43].

A major potential area for expansion in the most intensive forestry plantations is the use of biomass for energy. Woodfuel already supplies the primary energy source for about 2 billion people, mainly in the South. As governments start to address the problems created by future shortages of fossil fuels, interest in use of biomass is increasing, directly as firewood and also through various conversion stages such as use as a feedstock for gas and alcohol production. Major energy biomass plantations are being discussed in, for example, Sweden and the Baltic states.

Companies are anxious to present plantations as a sustainable alternative to logging native forests and stress the role of trees in carbon sequestration, watershed protection and as renewable energy sources.

CASE STUDY OF INTENSIVE FOREST MANAGEMENT: SWEDEN

Forest currently covers 58 per cent of Sweden, some 24 million ha, and is expanding[44]. The majority is dominated by Norway spruce (*Picea abies*) and Scots pine (*Pinus sylvestris*), although there are extensive deciduous woodlands in the south, and beyond the Arctic Circle conifers are gradually replaced by birch (*Betula* spp), aspen (*Populus tremela*) and willow (*Salix* spp). Swedish forests contain some of Europe's largest mammals, including bear, moose, reindeer and wolves.

However, all but a small proportion of Sweden's forests are now intensively managed, and replanted with single species after felling, as a result of state policy to encourage the forest industry. Perhaps nowhere else in the world has government and the timber trade worked so closely hand in hand in developing land-use policies. A critical element in this accord was the 1980 Forest Law, which virtually forced all landowners to fell their forests for timber. 'Protected' uplands were opened to exploitation in 1982 by the abolition of the voluntary 'forest cultivation border' rule which had protected trees on mountain slopes in western Lapland[45]. In 1983, legislation was extended to insist on clear-felling, on the grounds that opening the canopy allowed in light to stimulate regrowth. Other laws encouraged wetland drainage. Although this is now supposedly restricted, it is reportedly easy to get around controls, and, for example, swamp forests are still drained following clear-felling to allow re-establishment. In most forests today, forest management follows a regular cycle:

❑ clear-felling;
❑ replanting following site preparation by scarification, a maximum of three years after felling;
❑ thinning several times during the first year of growth to remove other tree species (pesticides are no longer used);
❑ application of nitrogen fertilizers 1 to 20 years before felling; and
❑ clear-felling when trees reach marketable size.

Natural old-growth forest halved in area during the 1980s and is now confined to a few, still-declining fragments; unless trends change there will be little left outside reserves in 15 years' time. Many protected areas are in the north and are treeless or only sparsely forested, while biologically important lowland productive forests, pine forests, northern spruce forests, coastal forests, woodland pasture and swamp forests are poorly protected[46]. Only 0.5 per cent of forest below the mountain areas is protected. Although government grants for establishing nature reserves exist, reserves are still being acquired very slowly[47].

Perhaps more significantly, Swedish forestry policy has altered the ecology of managed forests. There has been a general shift from deciduous or mixed forest to forest dominated by conifers. Spruce has been replaced by pine in the north, with a 20 to 30 per cent shift in the last 30 years, while conversely pine has been replaced by spruce in the south, where the latter grows more quickly. An increasing area is being planted with exotic species, such as lodgepole pine (*Pinus contorta*). Building forest roads has reduced wilderness areas and it is said that there are only four forest sites left in Sweden more than 10 km from a road[48].

These changes have had major impacts on biodiversity. Some birds dependent on climax forests are now declining, including the Siberian Tit (*Parus cinctus*). Half of Sweden's population of golden eagles (*Aquilia chrysaetos*) live in old Scots pine forest[49], which is declining. Many species of fungi, lichens and soil organisms also rely on climax forests, and are consequently used as indicators of old growth in northern Sweden[50]. Current conservation policy is now agreed to be insufficient to protect forest biodiversity. A 1991 report to the government suggests that 15 per cent of Sweden's forests need to be set aside for the preservation of biodiversity, but that this figure could be reduced if forest policy was improved, from an environmental perspective, on the remainder.

The extent to which deforestation has affected the Sami people of Lapland remains controversial. Old-growth forests contain epiphytic lichens that reindeer eat during the spring and autumn. Clearcuts produce hard-packed snow, making it more difficult for reindeer to dig up vegetation. The Sami now buy hay and deliver it with off-road vehicles, a change from traditional herding practice, although this is partly a consequence of larger herd sizes, both because in some areas there have been restrictions on the sale of meat introduced after fallout from the Chernobyl nuclear accident increased radiation beyond international safety limits, and also because of increasing intensification of reindeer herding.

It is argued that intensive forestry is necessary for Sweden's economy and for the rural workforce. However, despite a huge growth in forestry in recent years, employment has fallen dramatically because of mechanization and, most recently, the replacement of chainsaw operators by felling and extraction machines. According to recent statistics, forestry and associated industries employ only about 6 per cent of the workforce, with only 1 per cent in the forest itself[51]. While this makes forestry a major rural employer, the number of workers has fallen steeply in the last few years.

The real winners in Sweden's industrial forestry development have been the few large companies that control or have cutting rights in considerable areas of the forest estate. These include such giants as Svenska Cellulosa (1.7 million ha), Stora (1.4 million ha) and MoDo (630,000 ha). It is heartening that some of these companies are now addressing environmental issues, with, for example, Stora investing in research into more ecologically sustainable forestry methods (see Chapter 8).

CONCLUSIONS

This chapter has been critical of tree plantations and intensive forestry methods, as they are managed at present. Forest management and tree plantations have the potential for a positive environmental role in the future, both in terms of supplying timber, fibre and energy and as a way of taking pressure off remaining natural and semi-natural forests. It is disappointing that, at the moment, their contribution is all too often negative. Options for changing and improving forest management are addressed in Chapter 8.

6
Pulp, Paper and Pollution

Pulp, paper and board will soon consume over half the world's annual commercial cut of timber. Expansion in the industry means that timber is being used from previously untouched forests, along with fibre from intensively-managed, short-life plantations. Both these options have major environmental and social implications. Pulp processing is also one of the most polluting industries in many countries and paper use generates large amounts of waste. The impact of the pulp and paper industry is, therefore, of considerable importance to the development of forestry methods and to attitudes towards natural and managed forests.

This chapter concentrates on one particular use of trees: the pulp and paper trade. This is justified because of the scale of the pulp industry, which is expected soon to use up more than half the world's annual commercial cut of timber. Pulp production has impacts on natural and semi-natural forests and is the driving force behind the establishment of many tree plantations. It is also perhaps the form of timber use that leads to the most conspicuous consumption, and where disparities between rich and poor show up most starkly. Cutting across ecosystems, sectors of the timber trade and policy areas, the future of the pulp industry, more than any other single element, is likely to influence future plans regarding the world's forests.

GLOBAL PULP AND PAPER USE

The production of fibre for pulp, the processes of pulping and paper making, and the consumption of paper all take place mainly in the North. The production process is highly integrated, and there is comparatively little international trade in constituent products; indeed, some of the largest paper makers also consume the largest quantities of paper themselves. Both the sources and the users of paper are expanding however, and this may lead to greater cross-border trade. A recent price increase in paper will further confuse international markets.

World pulp production was 155 million tonnes in 1991, and 243 million tonnes of paper was produced[1]. Paper making accounts for roughly 1 per cent of global industrial output. Precise figures are difficult to calculate. Although the FAO publishes statistics for production of wood pulp and paper, paper can also be made from wood chips, wood residues, wastepaper, non-wood products, etc. On the other hand, not all these products are used for paper, some going to manufactured boards, etc. Over 35 per cent of world commercial timber production currently goes to pulp, and this is expected to increase to 50 per cent within the next few years[2]. Already more than 50 per cent of European timber is used for pulp[3]. Some countries, on the other hand, are moving away from wood fibre; for example Indian analysts calculate that by 2010 only 27 per cent of India's paper raw materials will be forest based[4].

Paper consumption is rising throughout the world. Since 1950, world consumption has increased five-fold[5]. World newsprint demand alone rose by 3.5 per cent in 1990, with above average growth rates in western Europe, Scandinavia and Asia[6]. The market continued to grow during the 1980s; for example between 1987 and 1988 the market for paper in North America increased by over 7 per cent, with similar rates of growth in many European states[7]. In 1992, International Paper, the world's largest paper company, had sales of over US $13.5 billion[8].

The North continues to dominate paper production, with over 80 per cent of output coming from the industrialized nations. Use is also far greater in the rich countries. Per capita use in North America is 60 times that in Africa and 150 times that in India[9].

Currently the most rapid rate of increase in pulp and paper use is in the so-called Newly Industrializing Countries such as South Korea, Taiwan, Singapore and Brazil. For example, in South Korea demand for imports and production of paper both rose by 12 to 13 per cent per annum towards the end of the 1980s, making it one of the world's most rapidly expanding pulp markets[10]. However, the continued and substantial dominance of paper making by the North makes these changes relatively insignificant in global terms. More than 20 African states have no commercial paper-making capacity at all. Elsewhere, there has been a (probably temporary) downturn in some former Soviet bloc countries such as Romania[11]. The pulp industry in former Yugoslavia is also in disarray[12]. On the other hand, pulp industries in Poland, the Czech Republic and Hungary have all developed since the introduction of a market economy[13].

The paper and board industry remains bullish about the potential for further expansion in the future. Use doubled worldwide between 1975 and 1991, and FAO predictions suggested a further 80 per cent increase by 2010. World capacity is projected to rise to 311 million tonnes by 1997[14].

Uses of paper are changing. In many countries over half of the paper use is for packaging and advertising and most goes to disposable or short-life products. Large quantities are also used in newspapers (where much of the space is for advertising), sanitary products and for office paper. In West Germany, in 1986 over 40 per cent of paper products were used in packaging, with roughly the same amount used for printing, including newsprint, and the rest divided fairly evenly between office and copy papers, tissues and hygiene

products and a variety of other speciality products. In the UK, it has been calculated that the paper thrown away every year is the equivalent of pulp from
some 130 million trees; more than two trees per person[15].

TRADE IN FIBRE, PULP AND PAPER PRODUCTS

Less than 20 per cent of wood pulp enters world trade, an indication of
the highly integrated nature of the industry. Pulp and paper-making mills tend
to be located near forests or plantations (or natural forests which are then
converted to plantations), in part because transport costs are a major proportion of total investment. However, trade does occur, particularly in fibre pulp.
This is imported by countries with a domestic paper-making industry but
without large enough supplies of timber, or perhaps more frequently because
cheaper supplies are available abroad. The main importers of fibre for
pulping are Japan, Finland and Sweden, together taking more than half the
total. These countries all have expansive forest cover and active domestic
industries, but can obtain a proportion of their timber more cheaply abroad.
Japan imports from a wide variety of sources on the Pacific Rim, ranging from
Alaska to New Zealand, while most Finnish and Swedish imports, including
fairly large amounts of birch, come from the Karelian area of Russia. Major
fibre exporters include the USA, the Russia Federation, Germany and
Australia.

More than half the world's exports of paper are from Canada (over a fifth
of the total), Finland, Sweden and the USA, although the USA is also a major
importer, due to a cross-border trade with Canada. Other significant
importers include Germany and Japan. In the South, the main exporter is
Brazil, and major importers include South Korea and China. There is a
growing trade from North America across the Pacific to Korea, China,
Taiwan and other Asian countries, some of which also import large amounts
of wastepaper for paper making. Other significant importers of wastepaper
are the Netherlands, Mexico and Canada. Chile supplies wood chips to Japan,
as do Australia and New Zealand[16].

ENVIRONMENTAL AND SOCIAL IMPACTS OF THE PULP AND PAPER TRADE

Paper and pulp production has impacts on forests, on other ecosystems
and on people in three main ways, through:

❑   timber (fibre) demand and the direct impact on forests;
❑   pollution and resource use during the pulping and paper-making
    processes; and
❑   the impacts of waste disposal.

Each of these is examined in more detail below, and the following case studies consider impacts in Japan and Germany.

TIMBER DEMAND AND DIRECT IMPACT ON FORESTS

The key role that pulp and paper plays in the international timber market means that it has a growing impact on both natural forests and secondary, managed forests and plantations. The large demand for fibre to create pulp has a number of ecological and social consequences:

❑   natural and old-growth forests are being felled to produce paper in areas such as the USA, Canada, Chile, the Russian Federation and in some limited but important areas of the tropics;
❑   forests, and other natural habitats, are being destroyed to establish plantations; and
❑   management of regrown or plantation forests is being intensified to meet higher demand.

At present, the bulk of pulp production occurs in a few major northern countries. The North continues to dominate paper production, with 1992 figures attributing over 70 per cent of output to the industrialized nations. The largest producers are the USA (31 per cent), Japan (11.5 per cent), China (8 per cent), Canada (7 per cent), Germany (5 per cent), Finland (4 per cent), Sweden (3.5 per cent), France (3 per cent), former USSR (3 per cent) and Brazil (2 per cent)[17].

In 1991, some 69 per cent of pulp came from coniferous timber. This proportion has remained stable for some time at a global level. However, several countries have added to the hardwood content of their pulp, such as Finland which uses an increasing amount of birch, and Australia and Portugal which have made use respectively of natural and planted eucalyptus. These changes have been balanced out worldwide by greater amounts of softwood becoming available from plantations, such as *Pinus radiata* in New Zealand and Chile. It is estimated that about 1 per cent of production comes from tropical hardwoods harvested in natural forests. The proportion from plantations is difficult to calculate, given problems in measuring quantities of plantations (and even in agreeing what constitutes a 'plantation'), but is likely to be between 15 and 30 per cent of the total.

However, some of the largest producers continue to fell natural or old-growth forests specifically for pulp production, for example in North America, Latin America, the Russian Federation, Asia, Africa and Australia. This has severe impacts on biodiversity, environment and, in some cases, indigenous peoples. Some important areas where natural forests continue to be destroyed to produce pulp and paper are:

❏ Pacific Northwest USA: old-growth forests are being felled for pulp and
   the new Republican administration has vowed to lift restrictions on
   logging, grazing and mining on public lands, greatly affecting remaining
   old-growth and roadless areas.
❏ Northern Canada: temperate and boreal forests are at risk from both
   Canadian and foreign timber companies clear-felling for quick profits,
   including forests under claim by native American groups.
❏ Western Australia and Tasmania: native eucalyptus forest is being felled
   and replaced with intensive monoculture plantations and there is an
   industry backlash against environmental protection.
❏ The Russian Federation: many native forests are under threat and timber
   from native birch forests is being imported into Scandinavia as a cheaper
   pulp feedstock than buying the home-grown alternative.
❏ Indonesia: over 2 million ha of tropical rainforest have already been felled
   and replaced by pulp plantations of acacia and eucalyptus and the government intends to cover 10 per cent of the land area with pulp
   plantations.
❏ Malaysia: at least three pulp mills using tropical hardwoods are either at
   planning stage or under construction.
❏ Vietnam: unique bamboo forests are being felled to supply pulp mills.
❏ India: felling of bamboo for pulp-making is considered to be an important
   environmental problem in some areas.
❏ Nigeria: there are proposals to use natural forests as feedstock for pulp
   mills.
❏ Cameroon: the world's first pulp mill using mixed tropical hardwoods was
   built in the country.

The industry is often reluctant to admit the importance of pulp in operations in natural forests, fearing resistance to logging purely for paper making.
In several areas where pulp is reported as being a 'by-product' of timber production, analysis has shown that it is actually the most important output from
concessions. For example, in New Zealand a proposed, and eventually unsuc-
cessful, beech forest scheme aimed at harvesting natural forest was presented
to the public for the production of high quality timber for furniture and
similar uses, with only a small wood-chipping component to use up waste
timber. However, the proposed sawnwood component fell away to just 20 per
cent in the first round of tendering, and it became clear that the final
proposals would essentially be for an industrial wood-chip scheme going to
pulp or other purposes[18].
    In Indonesia, felling for pulp is usually a two-stage process. Clear-felling
for pulp in industrial forest estates is only allowed if less than 20 m^3 of timber
exists per hectare. The strategy of the pulp industry has been to log natural
rainforest for timber selectively, ostensibly so that it can continue to be
managed on a regular basis, then to clear-fell once the quantity per hectare
has fallen below the permitted minimum[19]. Again, the intention to clear for

pulp is concealed for as long as possible. Towards the end of 1993, it was reported that Indah Kiat, a leading Indonesian pulp and paper manufacturer, had been fined Rupiah 1.36 billion (£400,000) by the Indonesian Forestry Ministry for alleged illegal tropical timber logging[20].

The natural-forest component in pulpwood is likely to continue for some time. In addition, many fast-growing plantations simply do not produce timber of sufficient quality to be used for anything but pulpwood, and these plantations are spreading throughout the heartland of pulp production in North America and Scandinavia. Manufacturers can also increasingly use short, five-year rotation plantations, grown with large inputs of agrochemicals, thus encouraging further demand for this type of forestry.

Pulp production encourages planting of fast-growing, exotic trees, including especially *Eucalyptus, Acacia* and conifers such as *Pinus radiata*. As described in Chapter 5, these can have major impacts on water systems, soil erosion, water acidification and spread of pests. In some cases, natural forests have been replaced by pulp plantations, as has occurred for example in Indonesia, Brazil and Chile. About 2.2 million ha of forest in Indonesia have been felled and planted with eucalyptus plantations and the government intends to convert another 6 million ha by the year 2000[21].

IMPACT THROUGH MANUFACTURE

Both pulping and bleaching of wood pulp cause severe water and air pollution problems. Paper manufacture is also resource intensive, requiring large amounts of energy and water.

Pulping

Pulping produces several pollutants, including organic products that cause eutrophication in water and aluminium salts. Some methods also produce air pollution, mainly from sulphur dioxide. A variety of methods are used to separate the cellulose fibres used to make paper from other material such as lignin and hemicelluloses. These are based on two main principles:

❑ **Mechanical pulping,** grinds debarked wood with a grindstone, or with rotating metal discs known as *refiners*. This is efficient in terms of conversion, but requires large inputs of energy. It produces pulp which is suitable only for newsprint, and other low-grade uses. Toxic wood chemicals left after mechanical pulping are often discharged.

❑ **Chemical pulping,** uses sulphur compounds to separate the pulp. This is less efficient, with only 45 to 50 per cent conversion, but produces a higher quality product. It operates a *closed loop* system. Waste material is burnt to power the system and many chemicals are re-used, thus reducing pollution. However, between 1 and 5 kg of sulphur dioxide are emitted for every tonne of pulp produced.

The two methods are sometimes used in combination, through *Chemo-Thermo-Mechanical Pulping* (CTMP). Here, sulphur softens the tissues before steam treatment and grinding. This reduces energy requirements but still results in significant toxic releases. Not all trees are suitable for pulping by all the methods available. A choice has to be made between efficiency of conversion, strength and quality of paper, amount and type of pollution, and resource use. Several less polluting production methods are being developed, including *solvo-pulping*, which uses alcohol to separate pulp from the lignin, but none of these are used on a large scale as yet.

By far the commonest pulping method used worldwide is the *kraft process*, a chemical method where wood chips are boiled in caustic soda. The resulting pulp makes high quality, strong paper, but is a dark brown colour and usually requires heavy bleaching[22]. A breakdown of pulping methods, and their associated pollutants, is given in Table 6.1.

Bleaching

To obtain pure white cellulose for paper, pulp has to be bleached. Traditionally this is carried out with chlorine gas, which is used to break down and remove the lignin, and chlorine dioxide or hypochlorite, which is used in a number of successive stages to bleach the pulp white. An average size pulp mill discharges some 30 to 80 tonnes of organochlorines a day.

POLLUTION FROM PULPING AND BLEACHING

Pulping and bleaching lead to the discharge of considerable quantities of water and air pollutants. Literally thousands of organic and inorganic compounds are discharged into wastewater, which usually enters river and lake systems, and many of these cannot easily be measured. The so-called 'conventional pollutants' include waste organic materials, many of which cause eutrophication. Wood fibres and bark threaten aquatic life if present in large quantities, sometimes by covering the bottom of rivers, lakes or shallow seas and smothering plants and animals.

More controversially, chlorine can produce a range of highly toxic organochlorine by-products, including dioxins, polychlorinated biphenyls (PCBs) and carbon tetrachloride. Up to 1000 different organochlorines can be formed during the process, although only about 300 have been identified to date. There is strong evidence that some dioxins are highly carcinogenic, including particularly 2,3,7,8-tetrachlorodibenzo-para-dioxin (TCDD). The US Environmental Protection Agency estimates that people who regularly eat fish caught near pulp mills have a thousand times greater chance of developing certain cancers than the average. Several areas of British Columbia have banned shellfish collection because of mill pollution[23]. In North America, both peregrine falcons (*Falco peregrinus*) and blue herons (*Ardea herodias*) have apparently suffered reproductive failure through accumulation of organochlorines from mill effluent. According to Environment Canada, during the winter period some 3 per cent of the water in the Fraser River – the largest salmon-spawning river in the country – is pulp-mill effluent and deformities have been found in fish near outfalls[24].

Air pollution is important in some mills, particularly older plant. Sulphur dioxide is the most important component of acid rain on a global basis, and chlorinated solvents released by chemical pulping contribute to both global warming and the breakdown of the ozone layer. Hydrogen sulphide, chloroform and carbon tetrachloride can all be released from mills and result in significant health risks to workers and nearby residents. The US Environmental Protection Agency has recently added the pulp and paper industry to its category of 'major sources of hazardous air pollutants' because of the presence of chlorine, other volatile organic compounds and chloroform in waste gases.

Workers at mills appear to face some direct and long-term risks from pollution. A study by the International Labour Organization (ILO) found wide differences in accident rates between countries, and some industry-specific diseases occurred. For example, in Finland a number of cancers were recorded at high incidence levels amongst workers in pulp mills, probably as a result of chemicals used in the pulp process and biological agents[25]. In the Russian Federation, the area around Bratsk in the east Siberian taiga was declared a disaster area in 1992, as a result of air pollution from the local pulp and board producer[26].

RESOURCE IMPLICATIONS

Paper making also uses large amounts of water, at rates of about 40 m^3/tonne in traditional mills, although this can be reduced through technical modifications. Annual use of water by the Californian Paper Board Corporation was 2,473,000 m^3; this fell to 689,000 m^3 following conservation measures[27].

TACKLING POLLUTION

Following a vigorous campaign about pulp effluent by organizations such as Greenpeace International, the pulp and paper industry in the North has reacted to reduce discharges. Some countries, notably in Scandinavia, and Germany, Canada and the USA, have introduced far tighter controls on pollution from pulp mills, and have already seen a dramatic reduction in pollution as a result. In Canada, Can $4 billion was invested in pollution equipment between 1989 and 1993, during a time of severe recession. Results can be dramatic. In Finland, for example, the Metsa Serla pulp mill near Äånekoski introduced pollution controls as a result of new Finnish regulations. Biochemical oxygen demand (BOD) loading of the plant's emissions into sur-

Table 6.1 Energy, resource use and pollution from pulping methods

Process	Efficiency	Quality	Energy use	Resource	Pollution and emissions
Chemical methods	45–50%	High	Low, as waste products are burnt to power the process	35,000–40,000 gallons of water used	Closed loop system, some pollution; cellulose fibres in wastewater can cause eutrophication; aluminium salts pollute water and kill fish; bad odour made Germany ban this method
Kraft process: boiling wood chips with caustic soda	45–50%	High, but needs bleaching	1150 kWh/tonne electrical energy; 5200 kWh/tonne steam energy	20 kg sodium sulphate/tonne, 75 kg calcium carbonate/tonne	1–3 kg SO_2/tonne
Sulphite pulp: boiling wood chips with sulphuric acid	45–50%	Fairly high, often used for tissue paper	1000 kWh/tonne electrical energy; 4400 kWh/tonne steam energy		5 kg sulphur dioxide/tonne
Mechanical methods	95%	Low: weak and discolours in sunlight	High, as no use of waste; 2000 kHw/tonne electrical energy	10,000–15,000 gallons water/tonne	Wood chemicals removed from pulp are routinely discharged
Thermo-mechanical pulping (TMP): wood chips softened by steam before grinding	95% ?	Stronger than simple mechanical pulping	Energy use significantly reduced as compared with mechanical pulping		As above
Chemo-thermo-mechanical pulping (CTMP), as above but first softened by sulphur	95% ?	Stronger pulp than TMP above			Routine effluent even more polluting through addition of sulphur; highly toxic and difficult to degrade

Source: Compiled by Equilibrium from data contained in R Kroesa (1990) *Greenpeace Guide to Paper* Greenpeace International, Vancouver; *The Sanitary Protection Scandal* (1989) Women's Environmental Network, London; *Dioxins* (1989) National Swedish Environmental Protection Board, etc. An earlier version of this table appeared in *Forests in Trouble* (1992) WWF International

rounding freshwater systems dropped from 50 tonnes per day to around 3 tonnes per day, which is less than the new legal minimum. Sulphur dioxide emissions into the atmosphere dropped from 11,000 tonnes per year to 200 tonnes per year[28]. Nearby waterways that were once almost completely dead once again have healthy fish populations.

Changes have taken place elsewhere as well. Dioxins were found in the effluent of two thirds of Japanese pulp and paper mills[29]. In October 1990, research at Ehime University revealed that fish caught in sea water near pulp mills in Iyo-Mishima contained 9.4 parts per thousand of dioxin. The Ministry of International Trade and Industry called for the 32 mills producing bleached kraft pulp by chlorine gas to change to oxygen or chlorine dioxide. There are now no mills in Japan using chlorine gas bleaching[30].

Despite substantial improvements in many areas, pollution still occurs. In the UK, 75 per cent of pulp mills exceeded their permitted pollution emissions at least once in 1990–92[31]. A Canadian Department of the Environment report of 1990 found 83 out of 122 pulp mills discharging waste above national standards. Many of the larger TNCs have initially responded to criticism about pollution with denial that there is any problem. Some have later worked hard to clean up their effluent problems, and the trade is taking an increasingly realistic attitude towards the inevitability of such changes. However, other companies have moved their operations to places where criticism will be less acute. Some plants, for example in former East Germany and the Russian Federation, have simply been closed as a result of pollution.

Problems are particularly severe in the former USSR. The Baikalsk Cellulose Paper Combine (BCPC) on Lake Baikal provides an extreme example. Work began on the mill in 1961, despite opposition on environmental grounds from a large group of scientists and academics. The 25 million-year-old lake is the world's largest and most ancient lake, containing 20 per cent of the earth's fresh water and about 1550 animal and 1085 plant species, including many endemic species. In its 30 years of operation, the mill has polluted some 200 km^3 of the lake's southern end, and dumped more than 900,000 tonnes of mineral salts and 18,000 tonnes of chlorides per annum. More than 4 million m^3 of lignin sludge has accumulated in sediment deposits. The communities on the lake bed have been destroyed over a length of 2.5 km and to a depth of 50 m. Atmospheric pollutants from the mill have destroyed 1273 ha of nearby forest and damaged a further 48,000 ha. Moves have now been made to include the lake and its watershed in UNESCO's World Heritage List, which would in theory ensure that there are no further industrial emissions into the lake. However, closing the mill has serious social implications, because 51 per cent of the male population of the 16,500 inhabitants of the town nearby are employed at BCPC[32].

Concern about dioxins has extended beyond the pulp-making process to the risks from residues remaining in paper and sanitary products, particularly in Europe, where scares about the impact of dioxins in coffee filters, disposable nappies and tampons have been the subject of much consumer concern. In Sweden, regulations have been tightened as a result of findings about dioxins in paper-waste effluent and alternative oxygen bleaching processes have been encouraged, so that by 1993 totally chlorine free (TCF) paper was available[33].

IMPACT FROM PULP AND PAPER WASTE

The disposable nature of most paper products means that paper is now a major, or even *the* major, component in many domestic waste streams.

Most paper is used for short-term or once-use purposes, and therefore ends up in the waste stream. Much waste ends in landfill sites. Some 40 per cent of rubbish dumps in the USA are said to be made up of unrotted paper. Apart from the space and health effects of landfill disposal, rotting paper releases methane, a greenhouse gas, which in addition can cause explosions if it builds up underground. Other paper waste is incinerated; sometimes energy from this is recovered and used for heating or electricity generation. Paper is not an ideal fuel and has been implicated in build-up of dioxin levels in cows' milk near incineration plants, although this issue remains highly contentious. Burning paper is nowhere near as energy efficient as burning the timber from which it is made. Paper can, in theory, be used in both composting and anaerobic digestion, although usually only as a portion of other biodegradable material and the amounts entering the waste stream tend to cut out this option in many cases[34].

RECYCLING

Most environmentalists' responses to pulp and paper problems have been to suggest reducing consumption and increasing recycling. While important, recycling itself has environmental side effects. Paper recycling can certainly offer substantial resource savings. The Warmer Campaign claims that replacing virgin pulp with recycled paper reduces industrial water use by 58 per cent, energy use by 40 per cent, air pollution by 74 per cent and water pollution by 35 per cent. Recovering the print-run of the Sunday edition of the *New York Times* would leave some 75,000 trees standing. Some 15 million tonnes of wood are thrown away every year worldwide in the form of disposable nappies[35].

The highest theoretical level of recycling possible has been calculated at 79.5 per cent, although the highest reached in practice is 53 per cent in the Netherlands. The amount recycled varies widely between countries; for example, New Zealand recycles only 16 per cent. There are some examples of very high rates of recycling. In the USA some cities, including Buffalo, Honolulu and Tampa, collect up to 80 per cent of office wastepaper for recycling[36].

However, recycling is failing to keep up with the rapid growth in paper use in many countries. In 1951, 32 per cent of paper used in the USA came from wastepaper, but this had dropped to 19 per cent in 1980 and, despite conservation efforts, is expected to reach only 21 per cent by the year 2000. Recycling is also being increasingly criticized, for pollution (through de-inking technologies, etc) and for the high energy use in collection, sorting and recycling paper[37]. These issues and questions gained official recognition in the UK, with criticism of high targets for recycled paper made by a Select Committee of the House of Lords[38]. Despite these reservations, most analysts still see an important role for recycling in reducing the overall environmental impacts of the pulp and paper trade in the future.

CASE STUDY: JAPAN

Imports and protectionism

Despite a large domestic forestry industry, Japan continues to import large quantities of timber for pulp. Japan is by far the largest importer of timber, by volume, in the world and, for example, imports 80 per cent of internationally traded wood chips. This situation is created by the high price of Japanese products, difficulties with the home industry and, to a lesser extent, environmental safeguards for forests. Accordingly, less than 50 per cent of the pulpwood consumption in Japan in 1992 came from domestic stocks. In 1991 major sources of wood pulp were (in order of importance) the USA, Canada, New Zealand, Brazil, Chile, Sweden, South Africa, Portugal and Finland[39]. In 1992 pulp imports totalled 2.9 million tonnes, about one fifth of total pulp production.

Japanese paper makers have been worried by growing concern for protection of forests, particularly in the USA, Canada and Australia. With this in mind, the industry is increasingly sourcing supplies from countries such as Thailand, Vietnam, the Pacific Islands, Chile, Western Australia, Brazil, South Africa and Argentina. Many companies are entering into joint-venture agreements with local companies to develop pulpwood plantations.

However, Japanese industry has been far less willing to import paper and paperboard, preferring to process fibre within the country. This has resulted in accusations of protectionism, particularly from the USA. Currently, paper and paperboard imports to Japan remain static at less than 4 per cent of total consumption. The Japanese government has responded to the criticism from the USA, and in 1992 a memo from Mr Kuriyama, Japan's ambassador to the USA, listed 16 measures by which the Japanese government could promote imports and 12 measures by which the US government could promote exports. In response to this initiative the National Paper Merchants Association of Japan and the Japan Paper Trade Association set up voluntary guidelines to boost foreign paper and paperboard imports.

Those imports which do occur are often from overseas companies in which Japanese firms have a vested interest. In 1992, for example, 14.7 per cent of newsprint consumption was met by imports. These imports were mainly from two mills: North Pacific Paper Corporation of the USA which is 20 per cent owned by Nippon Paper, and Howe Sound Pulp and Paper of Canada which is a fifty-fifty joint venture between Oji Paper and Canfor Ltd.

CASE STUDY: GERMANY

Resource conservation and pollution control

The demand for paper and board in Germany is high, with consumption in 1991 estimated to be 5,619,000 tonnes[40]. Production of wood for pulping is low, however, and the industry imports 64 per cent of its wood pulp. There is therefore a great incentive for local manufacturers to use wastepaper. In addition, the success of conservation groups in raising awareness of pollution issues has put the industry under severe pressure to reduce emissions and cut out chlorine.

Resource conservation

Wastepaper input into paper production has increased consistently and accounts for over 53 per cent of the total. Germany is also a large net exporter of wastepaper. Packaging papers and board consist of 93 per cent wastepaper, and newsprint usage of wastepaper, currently 72 per cent, is increasing.

Legislation has recently played a major role in encouraging waste management, with environmental legislation relating to recovery and utilization of recycled fibre. It includes:

❏ legislation concerning packaging: collection quotas of 30 per cent by 1993 and 80 per cent by mid-1995;
❏ draft legislation concerning newsprint: collection quota of 52 per cent;
❏ draft legislation concerning magazines: collection quota of 55 per cent;
❏ pending legislation concerning office paper: collection quota of 60 per cent by 1997[41].

In an attempt to reduce Germany's international forest impact, Greenpeace has been lobbying magazine publishers to use 'clear-cut free paper'. In December 1993, Greenpeace and four major paper consumers, including Mohndruck, the world's biggest offset-printing house and Otto, the world's largest mail-order house, produced a joint letter, condemning all destructive forest use and declaring their intention not to use paper produced from wood from destructive logging such as clearcutting. In early 1994, two major German publishers declared that they would no longer use any paper-based products from MacMillan Bloedel, because of its logging activities in Clayoquot Sound, Canada. In April 1994, the Association of German Magazine Publishers (VDZ) issued a statement that: 'the publishers will demand system-compatible forestry methods and the phasing out of destructive forestry methods and the phasing out of destructive forestry'[42].

Pollution control
The potential environmental and health aspects of dioxins caused much concern in the 1980s, leading to a wide uptake of TCF bleached paper in Germany. The kraft pulp production process is now banned, and the sulphite process has therefore become predominant. However, rising chemical costs and environmental concern have prompted investigation of new methods. One, *Organocell*, is a solvent-pumping process using sodium hydroxide, methanol and anthraquinone as the pulping agents. The resulting pulp is easier to bleach than kraft and is free from sulphur emissions. A large scale Organocell plant started up in 1992 but has since filed for insolvency[43].

CONCLUSIONS

The excesses of the pulp industry, at their most extreme, are depressing symbols of the worst aspects of timber use: intensive forestry, producing low quality fibre; polluting industry; conspicuous and largely unnecessary consumption; and problems with the waste that is produced. However, there is some good news as well. Developments in the pulp industry show that, when the decision is taken to change, that change can sometimes be very rapid. In the second part of this book (Chapters 7–10), potential solutions are considered, and it is worth bearing in mind the response of the pulp and paper industry to pressure about pollution.

7
Policy-based Solutions to Forestry Problems

As forest issues gain a higher profile, a series of national and international initiatives have been developed to address the problems. Unfortunately many of these are themselves deeply flawed and compromised. In this chapter we consider the role of non-governmental organizations (NGOs); reactions from the industry; international initiatives from the United Nations and multilateral aid agencies; and a series of national and international bodies and meetings, including the International Tropical Timber Organization (ITTO), Tropical Forests Action Programme (TFAP) and the various initiatives emerging from the Earth Summit (UNCED). The position of bans and boycotts is also briefly discussed.

If the world's forests are allowed to continue degenerating into the twenty-first century, it will not be for lack of international initiatives to secure their protection. Over the past decade, a bewildering array of trade agreements, multi-organization strategies, transregional plans, manifestos and conventions – to say nothing of countless workshops, symposia and conferences and the setting up of a host of NGOs – have focused global attention on the problems facing forests around the world.

Unfortunately, quantity does not invariably mean quality, as we have already argued in a different context in Chapter 2. The many different enterprises have all too frequently been poorly coordinated, at cross-purposes, or sometimes even in outright opposition to each other. Grand plans have been made without enough resources or the will to see them through, or drawn up so haphazardly that they have failed to work out in practice. Sometimes it seems as if words are used as an excuse for inaction. Problems that have delayed changes include political differences, sovereignty issues and the vested interests of an increasingly nomadic timber trade.

Now that there again seems to be a momentum building up for at least some level of change, many different governments, agencies and industrial interests are once more jockeying for position and dissipating a lot of energy in the process.

This chapter considers some of the policy responses to the issues raised in the first part of the book. It begins with a survey of the role and effectiveness of NGOs. This is a logical starting point, because there can be little doubt that the pressure of NGOs has been the spur to most environmental and social reforms of forestry, even if the reforms which do take place are seldom exactly what the NGOs had in mind. Next, some international activities are examined, including the role of United Nations bodies, conventions, multilateral development banks, aid policies and trade organizations. The importance of some recent events is then summarized, including the role of the Earth Summit and the Commission on Sustainable Development (CSD). Lastly, a few examples of the many national responses are briefly described, and put into an international context.

NON-GOVERNMENTAL ORGANIZATIONS

The international debate about forest conservation began by focusing on tropical moist forests. Given the huge scale of interest in forests among conservation organizations today, it is somewhat ironic that in the early 1980s most major environmental NGOs were very reluctant to become involved in the tropical forest debate. Indeed, in several countries small lobbying groups were set up out of frustration with inaction by larger NGOs. Staff at Friends of the Earth (FoE), the first international group to campaign against tropical forest destruction, initially argued that the subject was located too far from home, unlikely to catch the public imagination and so politically complex that any outside interference would simply lead to charges of neo-colonialism. WWF was so concerned that tropical forests would not be a popular campaign with members, that first forays into the area were deliberately presented as mainly concerned with primate conservation.

In fact, the public responded to information about threats to tropical forests with concern and enthusiasm. Most people proved more capable of grasping subtle environmental and political issues than they had been given credit for by the professionals. Tropical forests soon became a major pre-occupation for several large NGOs and, always an important consideration, also a project that was likely to be a successful hook for fundraising.

The role of international NGOs
Throughout the 1980s and in the 1990s, forests became an increasingly important issue for the NGOs. **Friends of the Earth** led the pack. It started in the mid-1980s with research linking UK timber companies with tropical deforestation[1]. It went on to investigate trade links throughout Europe[2] and as a result launched a campaign to boycott tropical timber[3]. While this was probably of more symbolic than economic importance, it was a significant warning to the European timber trade, which had at first been tempted to

ignore the letters and manifestos produced by FoE and its membership. FoE has since maintained an emphasis on research and has published, for instance, detailed monographs on specific aid projects, aspects of the timber trade and the role of tropical forests in climate change.

Next on the scene was the **World Wide Fund for Nature** (formerly the World Wildlife Fund). By the early 1980s, WWF had already been involved in running and funding field conservation projects for 20 years, including many in threatened forest areas. However, the organization had not, until then, taken a significant policy role, and had generally appeared reluctant to become embroiled in controversy and the 'environmental movement', considering itself a conservation rather than an environmental organization. New staff were prepared to take a more pro-active line with respect to forest policy, with the result that WWF gradually assumed a higher profile lobbying role, particularly with respect to the ITTO[4] and the TFAP (initially called the Tropical Forestry Action Plan)[5]. Partly through its monitoring activities of the ITTO, it also became a critic of the timber trade[6] and later set its own strategy and targets, and became heavily involved in the establishment of the Forest Stewardship Council (see Chapter 9).

In parallel with WWF, and with its international secretariat based in the same small Swiss town, Gland, the **International Union for the Conservation of Nature** (IUCN) has adopted a slightly more academic role in both policy and field projects. For the past decade, it has produced a steady stream of manuals and guidelines for management of tropical forests and of reserve areas, along with reports on specific projects within reserves, conference proceedings and its own strategies for conservation of tropical, and more recently also temperate and boreal, forests. Through the IUCN Conservation Monitoring Centre – which now has its own identity as the **World Conservation Monitoring Centre**, working from Cambridge, England – the organization has also played an important role in recording and mapping changes to the tropical forest estate, and the risks to particular species.

A more radical and people-orientated organization, the **World Rainforest Movement** (WRM), is based in Malaysia but has offices in the UK and elsewhere. Far less well funded, and consequently relatively limited in its scope of activities, WRM has been important as catalyst and critic. It has championed the rights of indigenous people and has been a particularly trenchant opponent of the TFAP. Opposition coordinated by WRM has helped distance other groups from the TFAP, and doubtless contributed to its virtual demise. WRM tends to be suspicious of large organized bodies and, for example, remains opposed to the idea of a global forest convention.

Other international groups have either had a more limited role or have been formed relatively recently. **Greenpeace** has been active on a national level, and has undertaken a few high-profile operations, for example in Siberia and British Columbia, but has not made forests one of its key international priorities. More recently, some international networks have been set up, including the **Taiga Rescue Network** in the boreal region and the **Native Forests Network** throughout the temperate countries. Activists and researchers increasingly communicate through electronic networking systems. Some international NGOs and their activities are summarized in Table 7.1.

Table 7.1 Some international NGOs active in forest issues

Name	*Notes*
Friends of the Earth (FoE)	Established a tropical forest campaign in 1984, linking UK and European timber companies and tropical deforestation. Organized a boycott against tropical hardwoods. Lobbies on trade and aid issues. Carried out work on illegal use of aid funding, the potential for sustainable management of tropical forests and the misuse of plantations as carbon stores
Greenpeace	Set up a forest campaign in the 1990s, although has worked on pulp and paper issues for years. Blockaded some illegal logging in Siberia and involved in campaign against logging of Clayoquot Sound in Vancouver Island, Canada
International Union for the Conservation of Nature (IUCN)	Research organization although also runs many field projects and has a policy brief. Involved in tropical forestry and has drawn up guidelines for management of reserves and for timber operations; now interested in all forests
Native Forests Network	Recently established international NGO, working on native forests and with a leaning towards direct action
Taiga Rescue Network	Networking body for groups working in boreal regions, linking organizations in Scandinavia and northern Europe, Russia and North America. Newsletter, meetings and occasional publications, including major report on the timber trade in the boreal zone
World Rainforest Movement (WRM)	Strong emphasis on local and indigenous peoples' rights and on sustainable use of the rainforest. Occasional publications and lobbying role; closely connected with *The Ecologist* magazine. Active critic of the TFAP
World Wide Fund for Nature (WWF)*	Has run field projects in forest conservation for 30 years; policy work began on tropical forest issues in the 1980s and extended to all forests in the early 1990s. Influential in promoting independent timber certification and establishing the Forest Stewardship Council. Many national organizations also involved in forest conservation

* in the USA and Canada is still known as the World Wildlife Fund

Source: Equilibrium (1994)

National NGOs and forest protection

Although the international organizations have the funds and the profile, in practice many of the most important activities have been undertaken by national or local organizations. These range from established and fairly orthodox conservation groups to colourful armies of villagers, tribespeople, anarchic groups of activists and sometimes new-age travellers who have mobilized, in different parts of the world, to protect forests from logging or management intensification.

It is fashionable to depict the environment as a predominantly middle class interest, dominated by the privileged and privately educated. Yet at the 'front line', the small organizations and groups that form when a particular habitat is threatened, show that this is far from the case. Generalizations are difficult. Responses to threats of logging or other industrial disturbance of forests remain varied and hard to classify. Sometimes local communities unite against the threats of forest destruction, while in other cases a variety of issues relating to profit, employment and land ownership cause splits within communities, or result in locals uniting against outside conservationists. Protests may be silent or vocal, passive or active, violent or non-violent. Examples of the range of local opposition to forestry developments include:

❑ Non-violent direct action against loggers by the Chipko movement in the Himalayan region of India, through villagers 'hugging trees' to prevent the use of chainsaws and further deforestation of steep slopes;

❑ use of legal procedures, including the Endangered Species Act as in the case of the northern spotted owl, to stop logging of old-growth forests in Oregon by the Oregon Natural Resources Council;

❑ letter-writing campaigns against companies operating in forest areas, such as those coordinated by the Rainforest Action Network in the USA and by Survival International in the UK;

❑ blockades of logging roads in Clayoquot Sound, British Columbia, to prevent logging of an intact watershed (over 700 people were arrested in the first half of 1994 during peaceful protests in the watershed);

❑ commissioning studies of the impacts of various forestry options on employment prospects in areas of the Pacific Northwest USA by the Wilderness Society;

❑ blockades of logging roads in Sarawak, Malaysia, by the Penan people and their supporters, protesting against logging of their traditional lands;

❑ use of a combination of activism, petitions, manifestos and political lobbying by the Beech Action Committee and other groups in New Zealand to prevent a major scheme for logging of natural southern beech forests in the South Island;

❑ non-violent direct action against road building through woodland areas in many parts of the UK, as in the case of Oxleas Wood;

❑ sit-ins and removal of timber from shops and timber yards in the UK, in

protest against the imports of illegally logged mahogany from Brazil;
❑ publication of joint manifestos by groups of NGOs involved in forest issues in places such as Australia, Denmark and the UK;
❑ lobbying by NGOs to secure voluntary boycotts on use of tropical hardwoods by many local authorities and corporations in Germany, the Netherlands and Austria;
❑ blockades of logging roads, backed up by use of legal actions through the Native Claims Settlements Act, by Native American groups in Quebec, Ontario and other provinces of Canada;
❑ successful lobbying for the creation of a World Heritage Area, and consequent declaration of a national park, in large areas of rainforest in Tasmania, by the Tasmanian Wilderness Society and others;
❑ legal action and union activism by Brazilian nut pickers and rubber tappers, to prevent further logging of areas of the Amazon in Brazil (the rubber tappers' case received worldwide recognition when their leader, Chico Mendes, was murdered in 1990).

NGO activity should not be put on a pedestal. At its worst, it can be poorly organized, oblivious to local people's needs, simplistic and divisive. More regularly, NGOs can be stubbornly, and sometimes needlessly, antagonistic, so that timber traders and conservationists are so totally in conflict that no attempt is made to seek compromise or middle ground. This situation has been common in, for example, parts of the USA. In these cases, even conservation 'victories' are tainted by the knowledge that they will be reversed by the industry wherever possible, and by a continued bitterness towards the environmental lobby. Of course, it is always easy to say how things could have been done better. Many NGOs have faced such entrenched opposition, to say nothing of corruption and sometimes violence, that 'reasonable debate' is often impossible.

None of these criticisms should disguise the fact that NGOs, both local and international, have frequently forced the debate about forest use. Changes in government and industry policy, which will be examined later in this chapter, have often been inspired as a direct response.

INTERNATIONAL RESPONSES

Over the past decade, a wide range of international responses have developed to address the problems identified and highlighted by the activities of NGOs, and given priority by the weight of public opinion around the world. A thorough review of these responses would require a weighty volume of its own. In the following section, a few of the key organizations and bodies are identified and their role outlined; some are summarized in Table 7.2.

Table 7.2 Major international attempts to control forestry

Name	Details
ITTA and ITTO	The International Tropical Timber Agreement, first agreed in 1983, and associated International Tropical Timber Organization; environmental credentials criticized ever since. Despite strong NGO lobbying, attempts to widen the agreement to all timbers were rebuffed in 1993 and 1994.
TFAP	The Tropical Forestry Action Plan, set up by the FAO, the UNDP, World Bank and WRI in 1985; ambitious plans to develop sustainable management for all tropical countries. However, a major impact was to encourage logging in previously untouched areas, and first WRI then the UNDP and World Bank withdrew. The TFAP, now renamed the Tropical Forests Action Programme, looks set to be abandoned at an international level, although bilateral agreements will continue.
CITES	The Convention on International Trade in Endangered Species of Plants and Animals helps monitor and control wildlife trade. It has taken some tentative steps in controlling trade in certain endangered tree species, and these have been mirrored by certain proposed legislation within the European Union.
Convention on Biological Diversity	Agreed at the Earth Summit (UNCED) in 1992; potentially important role in protecting forests, but has been weakened, its effectiveness not yet proven, and many countries have still not ratified the convention.
Climate Change Convention	Agreed at the Earth Summit; relatively minor role with respect to forest, although may have also a forest protocol relating to carbon sequestration. The aims of the convention were watered down at a Conference of the Parties in April 1995 as a result of pressure from the US government.
Agenda 21/Forest Principles	Some general principles agreed at the Earth Summit, mainly to placate those wanting to see a forest convention; now the basis of much discussion at the CSD.
CSD	Commission on Sustainable Development, set up after the Earth Summit and held a major meeting on forests in April 1995; this established an Intergovernmental Panel on Forests.

Source: Equilibrium (1994)

UNITED NATIONS BODIES

At least eight United Nations agencies or initiatives have had a direct impact on forestry and the timber trade, as outlined in Table 7.3. Of these, by far the most important to date is the FAO, with the UN Environment Programme (UNEP) and the follow-up from the UN Conference on Environment and Development (UNCED) potentially having a major role in the future.

Table 7.3 United Nations bodies and the timber trade

UN body	Full title and explanatory notes
FAO	Food and Agriculture Organization. Main UN body charged with forestry. Has tended to promote industrial forestry and, through the Tropical Forests Action Programme, opening up of tropical forests to exploitation. Now key agency (Forest Task Manager) in the CSD (see under UNCED below). Monitors tropical forest resources and publishes many forestry papers.
UNCED	UN Conference on Environment and Development, (the Earth Summit), Rio de Janeiro, June 1992. Failed to establish a global forest convention, but did agree some Forest Principles and was responsible for establishing the Commission on Sustainable Development (CSD), which includes a forest brief, and for agreeing the Convention on Biological Diversity and the Climate Change Treaty.
UNCTAD	UN Conference on Trade and Development. The body which originally coordinated negotiation of the International Tropical Timber Agreement.
UNCTC	UN Commission on Transnational Companies. Drew up guidelines for UNCED, with a timetable for implementation, for TNCs to reduce their environmental impacts. This was ignored and the secretariat disbanded. The UNCTC has moved from New York to Geneva.
UNDP	UN Development Programme. Influential in starting the Tropical Forests Action Programme, but later withdrew support.
UNECE	UN Economic Commission for Europe. Has a timber division which coordinates monitoring of temperate forests and of the timber trade, plays an active role in negotiations and treaties.
UNEP	UN Environment Programme. Establishes reserve areas around the world, funds small scale forestry projects.
UNESCO	UN's Environmental, Social and Cultural Organization. Establishes Biosphere Reserves, including several forest areas.

Source: Equilibrium (1994)

UN Food and Agriculture Organization

The FAO has long played an ambiguous role in the forest debate. It was one of the first international agencies to outline the threats to tropical rainforests through a series of reports in the late 1970s[7] and has started, funded and run many projects on small-scale forestry, non-wood products and related issues. However, its overall stance has been supportive of the concept of intensive, industrial forestry, and whatever the rhetoric at a particular time, FAO staff members have tended to view forests as timber resources rather than either ecosystems or multiple-use resources. Its funding, global reach, and close relationships with governments have allowed the agency to exercise enormous influence on the development of forest policy, and it has frequently been criticized for both its profligacy with money and its top down approach to development[8]. Today, the FAO has assumed the status of a bogeyman among many conservationists, culminating in a call for its closure by the influential magazine *The Ecologist*.

Tropical Forestry Action Plan

The centre-piece of the FAO's initial involvement in forest conservation was the Tropical Forestry Action Plan, which was launched with great fanfare in 1985, in cooperation with the UNEP, the World Bank and the Washington-based think tank, the World Resources Institute (WRI). Its stated aims were raising US $8 billion for tropical forest conservation between 1987 and 1991, setting up national forestry action plans, and providing an effective international umbrella organization for donor agencies, producer nations and NGOs. Edouard Saouma, then Director-General of the FAO, said that 'deforestation and forest degradation can be arrested and ultimately reversed'[9].

There were, initially, great hopes for the TFAP, and it won qualified support from environmental groups such as WWF. Unfortunately, the aims have not generally been achieved. The TFAP has failed to align donor agencies very effectively. It has not provided an integrated policy, nor produced a coherent set of country policies, some of which have actually resulted in *increasing* the level of timber production from primary forests. Studies have suggested that the TFAP has accelerated the rate of forest loss in some countries[10]. For example, in Cameroon, the National Forestry Action Plan in effect gave World Bank and UNDP backing to exploitation of the last 14 million ha of pristine rainforest in the south and east of the country[11]. Support for the TFAP, which was renamed the Tropical Forests Action Programme in 1993, has dwindled. In 1990, WWF concluded that fundamental changes were needed if the TFAP was to survive because:

TFAP has failed to provide consistent guidance and leadership to national governments in developing plans for the rational management and conservation of their tropical forests. It has been unable to set priorities amongst countries and consequently has become overburdened. It has yet to demonstrate the capacity to oversee the implementation of national plans. It has been unable to effectively align all donor agencies in the forest sector under one common umbrella. It has failed to ensure widespread participation by representative groups within countries. It has made limited attempts to integrate issues (such as land tenure, agricultural development, and rural poverty), which have a crucial bearing on forest sector development, into national forestry plans. And it has failed to convince local groups and non-governmental organizations that it offers realistic solutions over the long-term[12].

All significant environmental groups now oppose the TFAP, although the

International Institute for Environment and Development has remained more positive[13]. The other three original founder members, WRI, the World Bank and UNDP, have withdrawn support, leaving FAO holding an increasingly unwieldy baby. Indeed, WRI published a generally critical assessment in 1990[14].

Now, FAO is likely to have a leading role with respect to forests in the new CSD, described below, and most people believe that its support for the TFAP will gradually taper off to nothing. FAO also has a new Director-General and Assistant Director-General for Forests, both of whom are in theory committed to changing forest policy. However, it remains to be seen how far-reaching the changes are in practice.

Other United Nations agencies

Other agencies have had a direct effect on the trade. The ITTO (discussed below) originated through the workings of the UN Conference on Trade and Development (UNCTAD). UNESCO and UNEP both select, fund and sometimes administer forest reserves or areas suggested for special protection status. Through the Man and Biosphere Programme, UNESCO influences some key forest areas from an ecological and cultural standpoint, including for example Biosphere Reserves at Białowieża in Poland, Charkevoix in Quebec, Sinharaja in Sri Lanka, the Olympic Reserve in the USA, the Odaigahara Reserve in Japan and the Torneträsk Reserve in Sweden, many of which have at one time or another been scenes of conflict between the timber industry and conservationists.

The UN Commission on Transnational Corporations (UNCTC), although one of the smallest United Nations agencies, has played a potentially critical role with respect to the timber trade, through its work in drawing up environmental guidelines for TNCs[15]. Unfortunately, to date these efforts have been ignored.

The UN Economic Commission for Europe (UNECE), somewhat confusingly, monitors forest cover in much of the temperate world. The UNECE also publishes *Timber Trends*, a quarterly analysis of trends within the timber trade in the area, along with many papers on the timber trade in individual countries.

The United Nations is also the driving force behind many international environmental conventions and treaties, a number of which have a direct relevance to the timber trade. Some of the more significant of these emerged from the 1992 Earth Summit (UNCED), although older conventions could also become increasingly important in regulating trade in the future.

Perhaps the most important of the longer-standing conventions, in theory, is the **Convention on International Trade in Endangered Species of Plants and Animals** (CITES). CITES, also known as the Washington Convention, helps monitor and control the wildlife trade, and was originally set up with the aim of controlling trade in endangered species such as crocodilians, spotted and striped cats and marine turtles. It has also become important with respect to endangered plants, including many wild orchids, cacti and bulbs. More recently, the possibility of utilizing CITES in the monitoring of, and ultimately control of trade in, endangered tree species has also been recognized.

At present, 15 tree species are listed under CITES, most listings going back 20 years and all but one from Latin America. Listed species include Brazilian rosewood (*Dalbergia nigra*) and alerce (*Fitzroya cupressoides*) on Appendix 1 and the monkey puzzle tree (*Araucaria araucana*) and Mexican mahogany (*Swietania humilis*) on Appendix 2. Nine additional species were proposed for inclusion at the ninth Conference of the Parties in 1995, of which WWF backed inclusion of five:

- Agar wood (*Aquiliaria malaccensis*) from India, used in perfumes, incense, medicines and insect repellents;
- African cherry (*Prunus africana*), sold for its bark, timber and extracts used in medicines;
- red sandalwood (*Pterocarpus santalinus*) from the Indian subcontinent, used in food colouring, dyes, and furniture and musical instrument making;
- Brazilian mahogany (*Swietania macrophylla*), used in cabinet making, joinery, panelling, pianos, etc;
- Himalayan yew (*Taxus wallichiana*) from India, used in medicines, and for woodcarving and inlaying[16].

All but mahogany were listed on Appendix II. Whilst being an important step forward, none of these are traded for their timber, being used in pharmaceuticals and incense.

THE EARTH SUMMIT, THE COMMISSION ON SUSTAINABLE DEVELOPMENT AND OTHER INTERNATIONAL INITIATIVES

The United Nations once again took centre stage in the environment debate in June 1992, with the holding of the UN Conference on Environment and Development (UNCED), also known as the Earth Summit, in Rio de Janeiro, Brazil. UNCED was the 20-year follow-up to the highly influential Stockholm Conference on the Environment, which is viewed by many people as the launching pad for the modern environmental movement. It was billed as the 'last chance to save the Earth'. Given the bandwagon gathering behind the environment in the late 1980s, great hopes were held out for UNCED. The razzmatazz was certainly impressive. UNCED saw the greatest meeting of world leaders in history, and Rio was temporarily transformed, as environmental groups from all over the world gathered to lobby and demonstrate, and homeless people were quietly moved out of the city centre. Greenpeace activists floated banners accusing the whole process of being a sell-out, while the governor of the Amazon displayed opposing slogans demanding that all environmentalists went home.

In the event, UNCED was far from the turning point that had been hoped for. The results have been partial and, even worse, considerable ambiguity remains about the scope or remit of both the decisions taken and the follow-up activities.

Most of the main decisions were taken before the conference itself, including a complete rejection of the proposal for a global forest convention, initially vetoed by the Malaysian government but unmourned by many other states. In its place, UNCED produced a less binding 'Agenda 21' of plans for reducing problems of environment and development, and some 'Non Legally Binding Authoritative Statement of Principles for a Global Agreement on the Management, Conservation and Sustainable Development of All Types of Forest' usually known as the Forest Principles. At the conference itself, the signing of the Convention on Biological Diversity and the Climate Change Convention were the twin highlights, but three years on these both remain unproven and, in the case of many countries, unratified. The Biological Diversity Convention in particular, which had been viewed as a major plank in any post-UNCED conservation strategy by many NGOs, has been weakened through lobbying by the US government and is, in any case, woefully underfunded. In April 1995, the Climate Change Convention seemed to be in danger of disintegration in the face of refusals by the USA to meet targets for carbon dioxide reduction.

At the time, the rejection of a global forest convention was seen as a major set-back by many NGOs. However, the political row that erupted about the treatment of forests in Rio has, to some extent, meant that even greater efforts have been made than might have been the case if a flimsy convention had limped through the Earth Summit and then been ignored by many govern-ments, as is currently the case with the Biological Diversity Convention. Over the past two or three years there has been a flurry of activity as international bodies, research groups and NGOs try to interpret the UNCED decisions and apply them to international and regional policy.

The result is a series of different initiatives, many relating, directly or indi-rectly, to the UNCED follow-up meeting organized by the CSD in April 1995. In Table 7.4 some of the main activities are summarized. Many of these are, to one extent or another, attempts to define criteria for sustainable forest man-agement. To make up for lost time, many of the most strenuous efforts are being made in the temperate and boreal regions which, until recently, had virtually ignored these issues. Even for people who are closely connected with forest conservation, the bustle and commotion has been confusing. Different groups, with similar names and often great overlap in participants, have been meeting to try to decide policies relating to the same thing. Any country or organization without considerable staff and resources to devote to the task, risks simply being left behind.

Table 7.4 Attempts to define criteria for sustainable forestry

Name	*Details*
Helsinki Process	A joint initiative of the Finnish and Portuguese governments, launched with a European Ministerial Conference in June 1993 in Helsinki to agree four resolutions regarding forestry; now drawing up criteria for sustainable management.
Montreal Process	A workshop on sustainable management of temperate and boreal forests, run by the Conference on Security and Cooperation in Europe in Montreal in October 1993; now drawing up criteria for sustainable forest management among non-European temperate and boreal countries.
IWGGF: Malaysia-Canada	Intergovernmental Working Group on Global Forests set up by Canada and Malaysia; widely seen as an attempt to pre-empt, or control, moves towards a global forest convention. Produced a series of working papers as input to the 1995 CSD meeting.
Indo-UK Initiative	The UK and India drew up criteria for reporting on forestry-related issues at the CSD meeting in June 1995. Unfortunately, only 34 countries completed the surveys, and these were produced too late to be included in the FAO's Forest Task Manager report to the CSD
FAO: Ad Hoc Ministerial	The FAO is drawing up criteria of its own, although it is as yet unclear about how these would fit into the CSD process.
UN CSD	The Commission on Sustainable Development held a meeting focusing on forests in May 1995, which may act as a forum for drawing some initiatives in this table together.
CIFOR	The Centre for International Forestry Research in Indonesia is holding a Keystone International Dialogue on Sustainable Forest Management.
FSC	The Forest Stewardship Council has its own *Principles and Criteria* for good forest management which acts as a basis for any timber certification scheme affiliated to the FSC.
WWF	WWF has drawn up criteria for forest quality and has developed these in a series of workshops around the world.
National initiatives	Many national and regional governments are also drawing up criteria, including for example the USA, British Columbia and other provinces in Canada, Finland, and Sweden

Source: original list from J S Maini, Canadian Forest Service; details and additional initiatives added by the authors

Sometimes it is hard to believe that the confusion is not deliberate. In June 1993, a Council of Ministers' Meeting took place in Helsinki, Finland, to sign four resolutions regarding forests, as a follow-up to UNCED. The Helsinki meeting was jointly sponsored by the governments of Finland and Portugal, with the former very much taking a leading role. In September and October of the same year, the Conference on Security and Cooperation in Europe

(CSCE), an organization with very little track record on environmental issues, held a meeting in Montreal, sponsored by the Canadian government, billed as a 'Workshop on Development of Criteria for Sustainable Management in Temperate and Boreal Forests'. Following that, the secretariat of the Helsinki meeting, mysteriously transformed into the 'Helsinki Process', started to plan follow-up meetings and draw up its own criteria for sustainable forestry. The CSCE secretariat, in some documents also claiming to be a 'Helsinki Process' because the CSCE was originally set up in Helsinki, also planned developments along the same lines. Meanwhile, representatives from seven governments met in the Canadian embassy in Washington DC in December 1993 to plan their own follow-up to the CSCE, without informing other participants. Both the Helsinki and the Montreal conferences had separate meetings in Geneva in the same week in June 1994, the CSCE in some sort of cooperation with another Canadian initiative, this time in collaboration with the government of Malaysia. There are currently attempts to harmonize the various initiatives, started at a meeting at FAO headquarters in February 1995.

Multiply this several times, to account for all the other activities and ad hoc groupings, and there is more than a whiff of a turf battle in progress. FAO is pushing hard for a key role in future developments and is the Task Manager on Forests for the CSD, which gives it a leading position in negotiations with other international agencies. Unfortunately, despite a plethora of plans and proposals, it seems unlikely that many solid policy proposals will result from all this activity. The April 1995 CSD meeting was to have included the world's most detailed assessment of forest status ever, based on detailed country-by-country questionnaires. However, the reporting procedure finished later than scheduled and in the event only 34 countries completed surveys, and these arrived too late to be included in the FAO report. In the event, the April 1995 CSD meeting simply delayed an effective response to the demands made at UNCED, by establishing an Intergovernmental Panel on Forests to examine the issues more closely over a period of two years. The first meeting of the IPF took place in New York in September 1995.

The logical end-point of this wrangling would be some kind of global forest convention, and some environmentalist groups are calling for this. However, an increasing number of governments and organizations are worried that any such instrument would be both highjacked by the industry and implemented too late to help many of the most threatened forest ecosystems.

TRADE ORGANIZATIONS: THE ITTO

Stepping back in time a little, another critical initiative, which can be traced through the United Nations network but also had a wider genesis, was the **International Tropical Timber Agreement** (ITTA). This was first signed in 1985, and led to the setting up of the **International Tropical**

Timber Organization (ITTO) the following year. ITTO aimed to help research, regulate and coordinate the trade in tropical timber. From its inception it has been the scene of constant battles between the trade and the environmental movement, and has sometimes seemed to sum up all the differences and frustrations between the two points of view. ITTO was also the target for the most sustained lobbying by the world's most active environmental NGOs, which have provided staff, time and research in attempts at its reformation.

One structural problem with ITTO has been a series of differences in perception of its role, between producer countries, consuming nations, the timber trade and NGOs. Tropical-timber countries, major consumers of tropical hardwoods and the trade have all viewed ITTO as mainly a trade mechanism, which should be involved in *increasing* trade. Some consumer countries, and almost all environmental NGOs, hoped that ITTO would be a regulatory mechanism, for preventing the worst excesses of trade.

Largely as a result of constant pressure from NGOs, ITTO has taken some positive steps towards regulating the tropical timber trade, including setting a target that all internationally traded tropical timber should come from sustainably managed sources by the year 2000 and agreeing guidelines and criteria for sustainable forest management[17].

However, it has on the whole failed to implement its own policies at a domestic level. In Malaysia, the main focus of the ITTO's practical initiatives, the Sarawak state government has largely ignored the ITTO Mission's findings and recommendations which called for a drastic reduction in the rate of felling. ITTO has also tended to put great emphasis on projects, which have taken up a great deal of the organization's time in terms of approval and funding, and this dominated other issues[18]. In 1992, WWF concluded that: 'Regrettably, many ongoing ITTO projects must be regarded as subsidized logging, and the organization should ask itself whether it provides an incentive or disincentive for radically new approaches to forest management'[19]. The WWF review called for more emphasis on a 'low volume/high value' trade and for a more long-term look at the future of the tropical hardwood trade. It suggested national targets for sustainably produced timber volumes, rapid establishment of national guidelines for sustainable management of natural forests and a ban on log exports.

ITTO has also been a battleground between North and South which, in the current context of tropical timber, usually corresponds to consumers and producers. When the ITTA was up for renegotiation in the early 1990s, NGOs and later tropical governments such as Malaysia suggested expanding the agreement to include temperate and boreal timber as well. However, such moves were strongly resisted by the northern countries, which compounded frustration within the tropics by suggesting higher environmental standards for tropical countries than were simultaneously being agreed at the 1993 Council of Ministers' meeting in Helsinki for European forests, thus almost bringing the whole ITTA renegotiation process to a halt[20]. This marked a first major test of forest commitments made at UNCED, and the rich temperate and boreal countries were shown to be wanting. At the Helsinki meeting a single country, the Netherlands, suggested that tropical and temperate targets should be harmonized, but the suggestion was ignored. The renegotiated ITTA, finally agreed in January 1994, has, like the original, no environmental controls. By April 1995 only a handful of countries had ratified the ITTA, and the whole process appears to be grinding to a standstill.

MULTILATERAL DEVELOPMENT BANKS AND AID AGENCIES

Some of the most significant institutions and influences in the timber trade have been the various multilateral development banks (MDBs) and, to a greater or lesser extent, parallel funding from multilateral and bilateral aid agencies. Owing to the similarities between these institutions, all are discussed together in the paragraphs below.

Over the last ten years, the role of the various 'aid' institutions in forest development has been among the most contentious of all environmental issues. Starting from a position of being deeply committed to free trade, opening up markets and quick financial returns on investments, many of the lending and donor agencies have sponsored rapid utilization of forestry resources, through logging, development of roads, etc. Funding from many bilateral and multilateral schemes has increased net forest loss in places, through financing unsustainable logging operations, poorly designed hydroelectric projects and damaging road networks[21]. For example, analysis of World Bank projects undertaken in the late 1970s and 1980s found that many had a detrimental impact on tropical forests, including among others funding for:

- ❑ forestry development in Belize and Panama;
- ❑ management and exploitation of the Bosque del Novoriente national forest in Ecuador;
- ❑ development of logging in Guyana;
- ❑ management and intensive utilization of tropical forest in Peru;
- ❑ development of forest resources in Burkina Faso, Cameroon, Chad, Congo, Gabon, Nigeria and Senegal;
- ❑ access to the Mualnje forest in Malawi;
- ❑ forestry industry and technical support development in Mozambique;
- ❑ development of forestry in Bhutan;
- ❑ funding for paving of Highway 364 as part of the Brazilian Polonoroeste project to open up Rondônia;
- ❑ support for the Narmada Dam project in India which will flood huge areas of forest;
- ❑ development of the Carajas iron ore project in Brazil; and

❑ road building in forest areas of Vietnam, along with development of logging[22].

Criticism of the World Bank and other MDBs peaked during the 1980s, with campaigns run by the Natural Resources Defense Council and the Environmental Defense Fund in Washington DC, FoE and other bodies. Many new resolutions were passed within the agencies, and apparently real efforts made to address these issues by at least some staff. However, although there have been some improvements as a result, many national and international aid agencies have still failed adequately to vet what their spending achieves in practice. Nor do such agencies address the fundamental issues behind forest degradation, particularly in the South, as identified in Chapter 2. In spite of years of lobbying from NGOs, the World Bank's revamped forestry position, published in 1990, remains flawed and poorly implemented[23].

In recent years, unease about the World Bank's policy towards forests has continued. In 1990, the Bank put forward proposals for a US $23 million forestry and fishing scheme in Guinea, West Africa, including the felling of over 70,000 ha of forest. It was justified by claiming that the scheme would slow down logging which would take place in any case, and following criticism there are claims that a greater conservation component has been added. Site visits by environmental specialists reportedly found little evidence for changing in practice. Similar plans for funding Côte d'Ivoire's forestry sector were blocked because of the lack of any effective forest conservation strategy.

The World Bank has extended its influence within forests of the former Soviet bloc in recent years. There, mistakes made in the tropics look set to be repeated. For example, the World Bank has approved a loan of US $41.9 million for Belarus to establish forest development projects[24]. Environmentalists in Poland fear that a US $146 million loan will turn the country into a logging colony. A proposed loan for 'Ecological Management of Forests in Slovakia' threatens to double logging in a few years[25]. More recently, the European Bank for Reconstruction and Development (EBRD) has announced funding for the development of extractive forestry in the country, which threatens many of the surviving semi-natural forests.

NATIONAL INITIATIVES

Despite increasingly convoluted international mechanisms and proposals, a number of individual countries have taken steps to control the environmental and social effect of the timber trade. Such actions take two main forms:

❑ laws within a country to control aspects of forestry while allowing it to continue generally unabated; and
❑ laws or restrictions on forestry or transport of timber products, imposed either by producer or consumer countries.

Table 7.5 Bans, boycotts and restrictions on timber

Control	Explanation	Example
Total export ban	A legal ban on exporting any timber products from a producer country	Uganda
Partial export ban	Legal ban on exporting some timber products, eg not logs or some rare species	Ghana (ban on some rare species) Indonesia (ban on exporting logs)
Partial import ban	Legal ban on importing some timber products, eg tropical hardwoods, non-sustainable timber, etc	Proposals for bans on some tropical timber
Boycott	Voluntary boycott of certain timber products by people, local authorities and major purchasers	British Columbia timber, by some German pulp companies
Logging ban	Partial or total ban on logging, with or without parallel export restrictions	Thailand
Restrictions	Miscellaneous controls on rate of felling, etc	Brazil's timber quotas, many examples
Certification	Ban on importing or buying any timber that has not been independently certified as coming from a well-managed source	UK WWF 1995 Group (retail companies)

Source: based, with changes, on N Dudley (1991) *Importing Deforestation: Should Britain Ban the Import of Tropical Hardwoods* WWF UK, Godalming, Surrey

Within-country controls are, as might be imagined, extremely variable and often ignored in practice (see Chapter 5). More general restrictions on forestry and/or movement of timber fall into two main categories. A ban is a legally-enforceable ruling covering some or all species of timber, while a boycott is a voluntary rejection of certain timbers; some examples are given in Table 7.5. While boycotts, being voluntary, are impossible to counter by law, attempts at legally enforceable bans by either producer or consumer countries have often been challenged through free trade legislation such as GATT. For example, a ban on exporting raw timber from Indonesia has been challenged by the European Community and proposed bans on importing tropical timber by Austria and the Netherlands have also run into problems under international law.

Gradually, some countries are starting to work out and publish positions regarding domestic forest management, and this process should speed up in the wake of the Helsinki Process, CSD meetings and other international initiatives. However, it is still too early to know how far-seeing these will be or, the much bigger question, whether they will be implemented.

CONCLUSIONS

To date, international initiatives have failed to address adequately the problems outlined in the first part of this book.

8
Forest-based Solutions

Environmental and social problems of forestry can be tackled in a number of ways. The first is through changing forestry methods so that impacts are reduced. Contemporary knowledge of ecology is used to devise management systems that fit better into the framework of a natural ecosystem. The timber trade has mixed feelings about such approaches. On the one hand it offers a way of continuing in business, with, it is hoped less criticism from the environmental lobby. On the other, it risks reducing profits by swapping biomass harvesting for multipurpose forestry. Despite misgivings within the trade, a number of governments and timber TNCs are experimenting with a range of low impact forestry methods. Communities, individuals and aid programmes are also investigating small-scale forestry initiatives around the world. Many of these developments, while promising, are still too new to be assessed. Some traditional systems are also known to be less disruptive of natural ecosystems.

Most of the policy options discussed in Chapter 7 assume that forestry operations can be reformed in ways that lessen environmental damage. However, understanding of what might be called socially equitable and environmentally sustainable forestry remains fragmentary and incomplete. While it is fairly clear how non-timber benefits in some forest types can be maximized, in other forest systems sustainable management systems are still at an experimental stage, or have not yet been seriously considered. In this chapter, most of the discussion focuses on temperate and boreal forest areas, where the theory and application of low impact forestry has been receiving attention over the last few years. However, interest is also increasing, albeit slowly, among some sections of the tropical timber trade, and some tropical examples are also given.

APPROACHES TO FOREST MANAGEMENT

Over the past few decades, two distinct theories of forest management have developed:

❑ **Productive forests and reserves:** a policy of setting aside natural or near-natural forests as complete reserves while allowing highly industrialized forestry on the remaining area. This policy is based on the assumption that any management irrevocably alters forest ecosystems and that the only logical conservation response is to separate the two functions completely.

❑ **Multiple-purpose forests:** a policy assuming that almost all forests have already been altered by human activity and that management systems can be devised which allow timber extraction, other economic activities, environmental protection, biodiversity conservation and other functions to operate side by side in a given area of forest.

PRODUCTIVE FORESTS AND RESERVES

The case of New Zealand
In practice, the system of splitting forests into reserves and productive forests is really viable only in countries with a large surface area and a small population. It has gained particular significance in New Zealand. There, the need for exploitation of native forests has been removed by establishing large-scale, economically competitive plantations of fast-growing, exotic tree species. Forests are thus effectively divided into intensively managed timber production areas and untouched wilderness or other reserve areas.

Within New Zealand there is now a shared opinion among most foresters and conservationists that, broadly speaking, the correct choices have been made with regard to forest policy. This is not to say that there is universal harmony among the protagonists; indeed the conservation movement has undergone a bitter split over forest issues in the last few years. 'Choices' is also perhaps the wrong word; some of the developments have happened quite separately from each other and have only later been gathered together into an overall policy. However, it is significant that Greenpeace is the only major environmental organization actively campaigning on the issue of problems associated with plantation forests[1]. The largest conservation organization, Forest and Bird, even allows its name to be associated with timber from plantations, justifying this because plantations help protect native forests.

New Zealand's forests are important reservoirs of biodiversity. The country is one of the most significant relics of Gondwana, a prehistoric continent that included much of Australia and Antarctica, along with significant fragments of southern India, southern Africa, Madagascar and Latin America. Until the arrival of Polynesian and European settlers, life-forms evolved in isolation, with the result that New Zealand has an unusually large number of endemic species and genera. Flightless birds, such as kiwis (*Apteryx* spp.), occupy the ecological niche usually filled by herbivorous mammals. Some 65 tree and shrub species are endemic to New Zealand. Forests are dominated by either podocarp pines (*Podocarpus*) or southern beech (*Nothofagus*) along with various broadleaf species. Kauri (*Agathis australis*) is a dominant species in some northern forests[2].

It is estimated that forests originally covered 80 to 95 per cent of New Zealand, some 24 million ha. In the first 300 years of Polynesian settlement almost half the native forests, about 11 million ha, were destroyed. Serious European settlement started in the early decades of the nineteenth century and had two detrimental effects on remaining forests: continued and often accelerated felling, and introduction of large numbers of alien plant and animal species[3]. By 1975 it was estimated that 560 alien plant species had become established in the wild in New Zealand and 240 were common throughout the country[4].

Today, about a quarter of the original native forest remains, although some of this is regenerating following logging, particularly in kauri forest. The most recent detailed estimate was that about 6.5 million ha of native forest survive[5]. Many native forests have in the past been the subject of plans for logging, including a huge scheme for southern beech in the mid-1970s. Environmental groups resorted to lobbying, non-violent direct action, and long political campaigns to protect surviving native forests[6]. Ten years ago, it seemed almost certain that large additional areas of native forest would be felled, but this no longer appears likely.

Alongside native forest logging, the New Zealand government developed a large and intensively managed reserve of plantation timber, based mainly on *Pinus radiata*, the Monterey pine, originally from California. The 1920s and 1930s saw the first major planting boom, spurred on by predictions that native forests would be logged out completely within 40 years. From 1940 onwards there was sufficient plantation timber coming on stream to sustain a processing industry. There was a gradual switch away from using indigenous timber, to a reliance on exotic timber supplies. Exotic production first exceeded timber from native forests in 1955, and by 1990 indigenous timber had dwindled to a tiny part of a steadily increasing timber supply. New Zealand now supplies its own timber needs and is an exporter. Around 1.3 million ha of plantations have been established.

Development of a strategic reserve of plantation timber, which has now been privatized (described in Chapter 3), has helped take the pressure off native forests[7]. A series of hard-won conservation measures succeeded in protecting almost all remaining native forest and calculations in 1991 suggested that only 4 per cent was left open to exploitation[8], although some significant areas are at risk of disturbance and pollution from mining.

The general agreement within New Zealand was given a semi-official nature by the signing of the New Zealand Forest Accord in August 1991. The 14 signatories included four forestry bodies, eight conservation groups, including WWF New Zealand, the Federated Mountain Clubs and the Pacific Institute of Resource Management. The Accord includes the following objective:

> *To . . .recognise that commercial plantation forests of either introduced or indigenous species are an essential source of perpetually renewable fibre and energy offering an alternative to the depletion of natural forests...[9].*

Such a policy is an explicit rejection of the ideas of multiple-use forestry, which have been developed particularly in the USA over the past 20 years. It is grounded implicitly on the belief that ecologically sustainable forestry is impossible to practise in natural forest. It also accepts that intensive, chemically-based forestry practice is a necessary part of a country's forestry policy, so long as it frees land to conserve native woodlands in an untouched or old-growth state. Activists within New Zealand have argued that similar policies are needed in many tropical countries. This is at least tacitly the option being explored in countries like Indonesia and Malaysia.

Whatever the ecological problems associated with New Zealand's plantation policy, it is hard to argue that 1 million or so hectares of plantation is a bad exchange for 5 million ha of native woodland. Unfortunately, most countries do not have enough space to set aside such large reserves, or have already lost most of their natural and semi-natural forests. While reserves remain an important component of any national or regional conservation strategy, it is almost inevitable that a proportion of the managed forests will also have to supply other goods and services, including those relating to biodiversity, environmental protection, recreation, non-timber products, etc.

MULTIPLE-PURPOSE SUSTAINABLE FORESTRY

Multiple purpose forestry has, in theory, long been an important component of national strategies. However, as described in Chapter 5, in many forests the demands of the timber industry have virtually eliminated all other considerations, and other functions have been ignored or dealt with in a perfunctory manner. Now, increasing criticism of conventional forestry and of the sustained yield model is forcing forest managers to take the multiple purpose model more seriously. Among the responses, at a forestry level, are: demands for a return to more traditional forest management practices, especially in Europe; and calls for development of new practices, popularly known as 'new forest principles'. Multiple purpose management is not just about achieving ecological balance. It involves taking account of a wide range of social needs, relating to communities, visitors and the wider population. In some cases, the social framework of sustainable forestry is less well developed than the ecological theory.

The timber trade obviously has mixed feelings about both these approaches. On the one hand, threats of sweeping new conservation legislation, including the removal of additional forests from commercial exploitation, are causing ripples of concern through the industry in many countries. Any alternative is likely to get a hearing, and options of modifying existing practices seem more attractive than outright bans or designation of additional conservation areas. On the other hand, there is a deep conservatism within the industry, and fears that bringing on board 'new' forestry ideas will reduce efficiency, open up the forest to risks of disease and reduce competitiveness with regions where foresters face less stringent controls.

The issue has become highly charged in some countries. In the USA, where the old-growth debate is the chief domestic environmental issue for many people, there have been accusations that forestry policy is now effectively being set by politicians and ecologists, putting forward vague and unquantified plans in popular magazines. Some sections of the forestry industry have grasped at them relatively uncritically as a way of staving off further withdrawals of public land from logging. Professor Jerry Franklin, one of the forest ecologists most associated with new ecological principles of forest management, summed up the dilemma in an interview in 1990:

> *Basically people are interested in anything that even hints of a win-win solution to the conflicts we're in. Foresters are tired of being villains and the politicians are looking for some alternatives . . .there's just a logic about it. 'Hey, some of the things that we are doing just don't work ecologically.' And foresters are finally accepting just that*[10].

Political pressure doesn't necessarily help improve research in such sensitive areas as ecological land management. There is a real risk that systems will be developed and implemented too quickly, before they have been properly tested either for their effectiveness in protecting biodiversity or their silvicultural merits. Developments over the next few years will prove difficult and negotiations will sometimes be quite tense.

Basic principles of sustainable forestry
Despite the problems outlined above, some basic principles, and their practical management implications, are already starting to emerge from research. These include a range of ecological components:

❑ the importance of using natural species, local provenances and natural regeneration wherever possible;
❑ management practices which mimic natural dynamics including disturbance patterns;
❑ maintaining old-growth fragments as components of the system, ideally in locations where old forests would occur naturally in the landscape;
❑ relying on natural regeneration wherever possible, for at least a component of restocking after felling;
❑ planning for mixed forests;
❑ protection of particularly sensitive ecosystems within the forests, including

for example riparian (river and stream side) zones, wetlands, forests on steep slopes, etc;
❑ protection of key wildlife habitats, such as trees used as nesting sites for birds of prey, lek (display) areas for game birds, migration routes, etc;
❑ use of either very small clear-fells or selective logging; and
❑ keeping a proportion of standing dead timber (snags) and down logs in the forest, along with other dead wood components such as coarse woody debris.

In addition, various social components have been identified. While these are difficult to simplify, some basic principles include: recognition of user groups; security of user groups' authority to manage and utilize the products of community forests; equity of distribution of benefits; monitoring of performance and updating of operational plans; and technical advice[11].

Case study: New Forestry Principles in the Pacific Northwest

One of the best known attempts to put some of the ideas listed above into practice is the development of the so-called New Forestry Principles in the Pacific Northwest region of the USA. The principles consist of management changes aimed at reducing the differences between a managed forest and a natural-forest ecosystem. Their general philosophy is that forestry should, as far as possible, mimic natural disturbance patterns within a given ecosystem, thus attempting 'to balance commodity and ecological objectives by using practices based on current ecological understanding, the realistic limitations of the resource base, and societal expectations, including those expressed in Federal legislation'[12].

As formulated for the Pacific Northwest, the New Forestry Principles[13] start from the premise that current management strategies eliminate the early and late stages of forest development, and thus many of the species associated with old-growth forest. An 'uneven' management strategy may help overcome these problems. The general ideas have also been applied to forests in Sweden (see below). In both situations, they are designed for use in forests where fire is an important, and sometimes the most important, disturbance factor. Practical management implications include:

❑ leaving a proportion of trees standing after timber extraction, so that they can eventually develop into fully mature trees and thus maintain a varied-age stand[14];
❑ managing the amounts of coarse woody debris that enter streams, so that there is a sufficient quantity to supply additional habitats and nutrients for fish and other freshwater life, without completely blocking watercourses[15];
❑ planning for a mixture of tree species, including some broadleaved trees, usually by retaining a proportion of those which regenerate naturally[16];

❑ protecting trees growing in fire refugia along rivers and streams, ridges and wetlands, so that an interconnecting network of old growth can be established among more intensively managed forest areas[17]; and

❑ retaining a proportion of standing dead trees (snags) and down logs in managed forests.

Other factors have been suggested for some areas, such as:

❑ increasing prescribed burning (ie controlled burning of ground vegetation following felling), to mimic the impacts of fire and help retain those species requiring fire to survive;

❑ abandoning whole-tree harvesting to help retain nutrients within the system and reduce damage to soil structure;

❑ reducing or eliminating agrochemical use because of the side effects on wildlife[18];

❑ retaining wetlands and reducing drainage; and

❑ allowing some areas of forest to be dominated by broadleaved trees.

Because early descriptions were fairly general, New Forestry Principles as applied in the field can, and have, been used to describe a wide variety of management systems. These range from little more than cosmetic changes which disguise a 'business as usual' approach, to some genuine attempts at reform.

In a number of cases, forest managers have seized on the idea of 'mimicking natural disturbance' to justify almost any type of forest policy. For example, in the early 1990s there was a spate of claims that clear-felling in the boreal regions was equivalent to the effects of fire. This is untrue; most fires do not kill mature trees at all, and even the very occasional and severe wildfires usually leave standing trunks which provide habitat and shelter for seedling trees. The fire itself has an impact on soil and germination of seeds, which is missing from clear-felling. Such a simplistic interpretation has never been promoted by the originators of New Forestry, but has gained considerable currency amongst forest companies and is still regularly used to justify clear-felling policies in northern temperate and boreal forests.

Even in their original form, the ideas are controversial enough. Some foresters dismiss them as totally irrelevant, having been tried and failed in the past. In a paper delivered to the Oregon Society of American Foresters in 1990, William Atkinson delivered a bitter attack on the ideas of some of his colleagues at Oregon State University:

> *Most of us felt that nobody in their right mind would actually practice this stuff. But it turns out that not only are people applying New Forestry across the landscape, but proponents of New Forestry have been successful in catching the ears of some very influential people.*

He argues that New Forestry, which he calls 'hobby silviculture', fails to address issues of shortage of timber and land base and is technically a 'disaster' which will increase harvesting costs and risks while decreasing output. Major risks he identifies are:

❑ increased windblow from leaving individual trees standing, when they have previously grown up protected by an entire forest stand;
❑ escalating pest and disease attack from build-up in slash and transfer from old to young trees;
❑ greater potential for wildfire through leaving old timber in the forest as snags or down logs;
❑ soil compaction from multiple entry to carry out selective logging[19].

Many conservation groups are also distrustful about New Forestry, particularly in areas of old-growth or semi-natural forest. WWF US opposes its use in existing old-growth forests and argues that even less intensive forest managements will have detrimental effects on biodiversity in the few remaining old-growth forests in the country[20].

Obviously, the debate about forestry methods and low-impact forestry still has a long way to run. To some extent, the arguments are political; it depends by how much companies can be persuaded to reduce the amount of trees they harvest to satisfy non-timber benefits. These issues are decided by availability of resources, economics and pressure from the remainder of society.

COMPANIES GET INVOLVED IN NEW FORESTRY

In line with the greater role of private capital in forestry, some of the main research projects into New Forestry have been supported or run by major timber and forest-products companies, particularly but not exclusively in Scandinavia. There is a strong self-interest, of course, and sometimes the rhetoric runs faster than reality. In 1990, the Swedish firm MoDo was already claiming that:

When the trees are felled this is done with a view to preserving the character of the natural forest on each harvesting stand. In the 1990's harvested stands will not look the way they did in the past. They might look a bit more untidy because the dried out trees, fallen trees, tall stumps, bushes and undergrowth will not be removed. But the important point is to preserve the essential conditions for life for the flora and fauna and consequently the rich variety of species and diversity of the successively renewed forest[21].

Nonetheless, in the past five years some genuine and quite exciting attempts at changing management systems have been attempted. Most of these are in the temperate countries, although tropical examples are slowly starting to emerge.

Case study: New forestry in Scandinavia

For the last few decades, Scandinavian forestry has been systematically eliminating the oldest areas of woodland, and in consequence creating a forest landscape consisting of young, living trees. Wetlands, deciduous forest and old-growth trees have all but disappeared in some areas. The ecological implications of this are gradually being recognised. For example, the value of older trees, and of dead and decaying timber, has been reassessed. Over a thousand Swedish beetle species depend on dead trees. On spruce, 64 species of insects are specifically associated with living trees, while over 300 species rely on dead timber[22]. A survey of a single dead stump over one growing season found over 20 red list insect species.

Since the Earth Summit in 1992, the Scandinavian forest industry has made a series of attempts to address issues of forest quality, spurred on at least in part by NGO pressure both within and outside the region. Both Sweden and Finland have passed important new forestry laws since 1992. The industry now claims that 'the goal for Scandinavian forestry is to preserve biodiversity'[23]. Whilst some of the earlier statements from the timber trade tended to stress past achievements, and particularly the reforestation of Fennoscandia during the twentieth century, more recently, the emphasis has increasingly been on changes in contemporary forest policy, to give greater opportunities for biodiversity and the maintenance of natural ecological cycles[24].

Several of the larger forest companies now employ professional staff specifically to address issues of biodiversity and forestry. At Stora Skog, the large Swedish forest company, ecologist Börje Pettersson has been introducing methods of ecological landscape planning, developing forestry techniques for the protection of the rare white-backed woodpecker and experimenting with forest management methods aimed specifically at imitating natural processes within the forest[25]. Prescribed burning has been reintroduced in areas that naturally experience regular light fires, to mimic a fire ecology. More of the forest is left unmanaged, including naturally non-productive areas, rock outcrops, small forests in bogs, bog edges and other divergent biotopes. A proportion of waterlogged broadleaved and coniferous forests are also retained, while forests on steep hillsides are managed with a variety of protection zones. Ravines in sedimentary soil types are left entirely intact. Standing dead trees, down logs and some living trees are preserved after felling, along with a few high stumps as nests for tits (*Parus* spp.) and other birds. Bare logging areas larger than 3 ha are broken up by leaving at least 10 trees in groups. The proportion of broadleaved trees in the landscape is being gradually increased. After management, it is aimed to have at least 20 per cent hardwoods in spruce management and 10 per cent in pine management areas. Key biotopes, including habitat of red-listed species (ie those identified as being at particular risk of extinction) are also left outside management. Cultural remains, including ancient monuments, traditional paths, abandoned hay fields, old forest grazing lands and derelict settlements are also left. In 1995, for the first time, the company produced a 'green balance sheet', an internal appraisal achievement of forestry and nature conservation objectives[26].

Other Swedish companies are also developing new forestry methods, including AssiDomän[27] and Korsnäs[28]. Both these companies have also produced ecological balance sheets and plans for future forestry management. For example, AssiDomän's landscape planning will entail 8 stages:

❑ Identification of areas for planning, ideally delineated by natural boundaries such as watersheds. Average areas will be 5000-10,000 ha in the south and up to 25,000 ha in the north.
❑ Existing valuable habitats will be located in these areas, on both company and private land, over a period of 5 years.
❑ Natural watersheds will also be identified and mapped.
❑ Dispersal corridors will be identified.
❑ Natural disturbance patterns will be identified in the various areas and their imitation included in forestry planning.
❑ The proportion of broadleaved trees in the landscape will be measured, and goals for the future percentage identified and set;
❑ The need for extended rotations will be set;
❑ The need for reintroduction of fire will also be assessed, along with the location and ideal frequency of such fires.

On AssiDomän's holdings, 15 per cent of the land on average is preserved from management. A landscape plan should be in place on all the holdings within 5-10 years[29].

There has, undoubtedly, been a revolution in forest management within Scandinavia over the last few years, at least in public statements and plans regarding forestry. It is still too early to see how these ideas work out in practice. Whilst some of the theoretical costs can be avoided, because the areas left unmanaged tend to be those that are more costly to fell and replant, there undoubtedly is an additional cost for the producer. The extent to which this can be maintained in the face of cheaper alternatives from less well managed forests has yet to be tested on the marketplace. The industry's commitment also still falls short of allowing independent verification of good practice, although there are strong signs that Sweden will soon agree a national standard for timber certification. The Finnish timber industry, which has been fighting similar attempts to introduce timber certification in Finland, is thought likely to capitulate and do the same in the face of a positive Swedish decision. Events over the next few years will determine whether the various companies are involved in a complicated public relations exercise, introducing well-meaning plans that are later abandoned, or are really changing the fundamental philosophy of forestry in the region. If the latter is true, it could have implications for forest practice throughout the world.

Although the majority of work has taken place in the temperate and boreal zones, some work has also been carried out in the tropics, generally relating to logging practice.

It is, however, still too early to assess results very fully. In Sabah, Malaysia, experiments are underway into various options for reduced impact logging (RIL) which aims to produce a logging system with a 50 per cent reduction in damage to both the soil surface and residual trees. Basic processes involved include:

❏ a complete inventory and mapping of actual and potential crop trees;
❏ designation and mapping of rivers, buffer zones and fruit trees;
❏ cutting canopy vines and creepers at least six months before felling operations take place;
❏ marking trees for felling (with diameters above 60 cm) along with potential crop trees (anything with a diameter above 10 cm);
❏ planning of skid trails and tractor haul trails to minimize the area affected; and
❏ introduction of methods for protection of surface soil and drainage.

Detailed guidelines are being produced for road construction, stream crossing, skid trails, etc and there are plans to add the retention of wildlife trees. However, the commercial implications of RIL are significantly higher costs, in part because fewer trees are taken, with 15 trees/ha taken in conventional logging as compared to only 10 to 12 trees/ha in RIL systems.[30]

CONCLUSIONS

Changing forest management policy can help reduce many of the detrimental social and environmental impacts currently associated with forestry operations. Nonetheless, there are some limitations to this approach:

❑ however well forests are managed, they cannot fulfil all the ecological and biodiversity functions of natural or semi-natural forests, so that management changes supplement, rather than replace, the need for a good protected-areas network;

❑ there is no universal blueprint for management changes: although general principles exist they must be adapted and tailored to individual ecological and social conditions; and

❑ if industry is to be persuaded to take these issues seriously, some way of checking and verifying that management changes have taken place is required.

The last point is addressed in Chapter 9, which looks at market-based solutions to forest management problems.

9
Market-based Solutions

Policy changes and experiments with forest management will have little overall effect unless the timber trade itself can be persuaded of the need for reform. One way in which this can occur is through pressure on the market from NGOs, consumers, etc. The timber trade has initially responded to such pressure with increasing use of many labels and by claiming that timber comes from well-managed sources, but without any means of proof or verification. Disquiet with such a laissez faire *system has persuaded interests in the NGO community and the timber trade to promote independent certification of timber and establish the Forest Stewardship Council (FSC) to oversee this process.*

So far, our discussions about reforming the timber trade have tended to follow traditional lines: NGOs lobby for policy to be set by national and international agencies, and in consequence forest managers change their practices to reflect new priorities. These mechanisms will, certainly, continue to be important in the future. However, there is a 'third way' of changing the direction of forest management. Increasingly, policy is being reformed from *within* the timber trade, either because of pressure from NGOs, consumers and shareholders, or because of changing perceptions within the trade itself.

These **market-based solutions** will become increasingly important in the future. They allow consumers to interact directly with producers in setting priorities for the production and management of forest products. Such mechanisms are part of a long tradition of consumer pressure influencing the marketplace. Examples include voluntary boycotts on whale products and tuna caught in ways which harm dolphins, demands for additive-free foods and the long-running campaign against baby milk products in the South. Currently, considerable efforts are being made to develop mechanisms, and an overall framework, which will allow such concerns as they relate to forest products to be expressed, and the responses monitored and reported.

The significance of consumer influence is variable and liable to over-simplification or exaggeration. In many cases industry has, on the contrary, been very successful in influencing consumers with respect to what they think they want. Indeed, persuasion of the consumer that a particular product is worth buying is the basis of much of the current high consumption economy.

Industry therefore tends to be ambivalent about the reverse phenomenon, sometimes attaching great significance to quite minor manifestations of discontent among potential purchasers, while on other occasions apparently ignoring or underestimating real levels of concern among a significant proportion of consumers.

The extent of changes within the timber trade should in consequence not be exaggerated. As discussed below, the first instinctive response to green consumerism from a large section of the timber trade has either been to ignore it altogether or to attempt what has become known as a 'greenwash', that is, to claim that products are produced in an environmentally responsible way, without providing much in the way of details or proof. (Other responses have been to attack and attempt to discredit NGOs, see Chapter 7.) Such tactics have been successful in buying time for industry over the past five or ten years, but are now coming under attack.

There have, nonetheless, been some more radical developments within the industry with regard to changing policy in response to consumer demands. (The response within the industry has been touched upon in Chapter 8.) Indeed, when the pressure is strong enough, and the case fairly clear-cut, industry can often move faster than governments (and sometimes than many NGOs). The speed of the response in many countries to fears about pulp and paper pollution has been impressive, although it should be remembered that this came at the end of a long period of campaigning, boycotts and government pressure. It may be that we are now in the midst of a similar bandwagon of change regarding forest management. The signs suggest that this might be the case, although it is still too early to know whether this will have much real impact, and support remains tenuous. At the moment, many developments hang on the decisions made by a few fairly major figures in the timber trade, some of whom have a strong personal commitment to reform. Changes in economy, personnel or political climate could still put many of these initiatives at risk.

In this chapter, we look first at attempts by the industry to respond to consumer pressure without actually changing practices very much. Then, attempts to monitor and certify changes within the timber trade are described, including a summary of progress with the Forest Stewardship Council (FSC). Finally, some examples of market-based approaches are described.

The setting up of the FSC has been a particular focus for WWF and for the authors of this book. It is, therefore, treated in more detail than many other issues tackled herein. Nonetheless, our thesis is not that timber certification offers any kind of universal panacea, and the actual and potential shortcomings of the FSC are noted, and a discussion of its options and limitations is given in a concluding section.

CLAIMS OF ENVIRONMENTAL ACCEPTABILITY

The rise of green consumerism in northern countries over the past two decades has encouraged producers, traders and retailers of wood and wood products to respond to customer demand with a rash of unverified claims of 'environmental friendliness' for their products.

Although the majority of *tropical* timber-producing countries have made commitments to sustainably manage their forests by the year 2000, similar commitments have not yet been forthcoming from temperate and boreal forest producers, with the exception of the USA and Australia. Some governments, such as the UK, believe that their forests are already managed sustainably, although there have been some important international decisions to *introduce* sustainable management which suggest that the shortcoming of many systems is being recognised.

Over the past few years, publicity about deforestation and forest problems has led to a vast array of claims from within the industry. Initially such claims might have been simply that products contain no tropical timbers (spurred on by attempts to boycott tropical hardwood imports and use), or more recently that timber comes from well-managed resources. It has become difficult to buy a timber or paper product *without* such a claim in some countries.

These claims are worth making. A MORI poll carried out for WWF UK in April 1990 found that 25 per cent of people would be prepared to 'stop buying wood products made from trees such as teak or mahogany unless it could be guaranteed that they come from countries that were protecting their forests'. This survey was carried out before a high profile series of campaigns against the use of these timbers by Greenpeace, EarthFirst! and the Rainforest Action Group, which might be expected to increase public awareness of the issues. Many consumers are, therefore, almost certainly choosing products because they claim to have no damaging effects on the world's forests.

Since 1990, WWF UK has monitored and investigated the reliability of such claims, first for imported tropical timbers and more recently for all timber products. Two major research projects have found that virtually *none* of the claims that timber comes from sustainably managed forests can be verified, and many products, on the contrary, clearly come from forests that are being unsustainably logged or badly managed. Out of an initial survey of 626 companies, only three were found which were prepared to make a serious attempt to justify their claims, and none could be considered to have fully answered all questions regarding their sources.

There was also considerable confusion among members of the timber trade about the origin of their products, what constituted good forest management and the likely impact of the trade. Many interviewees believed that replanting was a sufficient guarantee that environmental issues were being adequately addressed, and confused 'sustained yield' (see Chapter 5) for 'sustainability'. Many traders believed that issues of concern regarding forest management were confined to tropical rainforests. Companies relied heavily on information given to them by suppliers and trade bodies without bothering to check it too carefully. Although there was a consistent and clear wish to present a 'green image' to the consumer, there was little desire, or knowledge of how, to back up claims[1].

Instead, the timber trade has, in effect, taken to self regulation in an attempt to regain the moral high ground. Both governments and industry, particularly in Scandinavia and North America, now engage in self-certification with apparent impunity, claiming that their products are 'from a sustainable source', 'from a reforested area', 'from forests that are sustainably managed', 'from environmentally managed and sustainable forests', or that 'for every tree cut down we plant four new ones'. These kinds of claims now accompany the whole range of wood and wood products, from timber and furniture to greetings cards, food cartons, personal hygiene products and stationery. Most mean little in practice. As interest in temperate forests has increased, so have the claims about pulp and paper, but again these are seldom substantiated. In some countries, for example, paper products can be labelled as *recycled paper* even if they contain only a small proportion of recycled fibre, and the same symbol can be used for recycled as *recyclable* paper, further confusing the purchaser who does not stand in the shop examining each label with minute attention.

In January 1993, as a test case, WWF UK referred a claim by NHG Timber Ltd, which had advertised 'environmentally acceptable Redwood' produced from 'sustainable forests', to the UK Advertising Standards Authority (ASA). The ASA upheld the complaint that no proof was offered for the claim. Six months later, the ASA also upheld a complaint against the Malaysian Timber Industry Board, which had advertised that its timber was 'from one of the world's best conserved forests'. In February 1994, the ASA upheld a third complaint, against Magnet Trade which is one of the UK's largest suppliers of doors, windows, kitchens and bedrooms, that 'where hardwood is used, it has to come from a renewable source'. Most recently, a complaint has been referred to the ASA by the Temperate and Boreal Forest Campaign, regarding the Finnish timber industry's use of the *Plusforest* logo, accompanied by claims of 'sustained forest' and 'green excellence'. In a number of other cases, where complaints have been drafted against companies, the claims have been voluntarily withdrawn rather than have the case referred to the ASA.

It is simplistic to see conspiracy everywhere. Deliberate fraud does exist, and on a massive scale in some countries as we have already described in Chapter 4, but ignorance is an important factor here as well. In many cases, companies (particularly those at the end of the purchasing chain) are simply unaware of any problems with their products. Self regulation within the industry has allowed a situation to develop in which fraud, semi-fraud and straightforward confusion combine to create a state where pressure from consumers is being countered by a string of deliberately misleading, or at best naïve and disingenuous, claims from producers.

This is not entirely the fault of the industry. Recognition of forest problems has come fairly late, and knowledge of how they should be addressed remains sketchy. Paper makers felt that by cleaning up pollution associated with paper manufacture they had met demands from environmentalists, only to find their forest management policies the target of fresh attacks. Principles of good forest management exist (such as those of the FSC), however at present detailed knowledge about how to manage specific forest types is still lacking. As a result, no blueprint can be offered for improvement. Tackling the issues, rather than simply criticizing the results, is inevitably a messier and less clear-cut process.

THE MOVE TOWARDS TIMBER CERTIFICATION

The purchasing policies of companies, local authorities, architects and specifiers have a direct impact on the world's forests. A weak environmental purchasing policy (or lack of policy) creates an incentive for forest producers to supply timber at the lowest possible price, with disregard for the environmental and social consequences, thereby contributing to the destruction or degradation of a forest somewhere in the world. However, a strong purchasing policy can stimulate a strong incentive for improved forest management, which will in turn ensure that suppliers conform to internationally agreed standards of forest management, designed to include the social and environmental costs of logging. In 1994, WWF argued that: 'Companies and individuals who continue to buy cheap timber on the open market risk buying into the process of deforestation and forest degradation worldwide, some of which is being caused by the timber industry'.

In 1990, as a result of the various issues summarized in this book, pressure began to build up for a system of *independent timber certification*, to bring order to a confused and confusing marketplace. Under an independent system, companies claiming to produce or purchase timber from well-managed sources have to prove this, and are provided with the framework to enable them to achieve such proof. Following lobbying from NGOs, the World Bank stressed the need to establish systems of timber 'ecolabelling' when it published its new *Forest Policy* in 1991.

Potential advantages of timber certification
Timber certification, at its most ideal, offers a number of distinct advantages:

❑ it helps distinguish those companies making genuine efforts at reform from those making cosmetic or fraudulent efforts, thus allowing differences in market penetration and price to reflect real changes in the forest;
❑ it provides companies with a clear idea of what is expected from them in terms of forest management, buying policy, etc, and some assurance that if they meet the required standards they will avoid further stressful and

damaging conflicts with NGOs and others;
❑   it provides an open forum for industry and other interests groups to debate issues regarding forest policy and management; and
❑   it gives positive incentives for change.

However, note the word 'ideal' in the introduction to the list above. Timber certification also has the capacity to disappoint rather badly if it fails to carry the industry and NGO community with it, or fails to provide real teeth for regulation of the markets.

Development of the idea of certification

A number of stages in the process of recognizing the need for certification can be identified:

❑   criticism of existing forestry practices and timber sources by NGOs and concerned individuals;
❑   increased consumer demand for alternatives to products from badly managed forests;
❑   establishment of companies dealing only in 'well-managed timber', which they selected with care from known sources;
❑   appearance of NGOs, companies and groups interested in acting as certifiers for timber products; and
❑   development of the concept that such schemes needed some established overseeing body to satisfy customers, and the timber trade, that claims were accurate and fell into a generally acceptable framework.

In the late 1980s, a small number of companies were set up deliberately to deal in timber from well-managed sources. In the UK, the Ecological Trading Company, based in Newcastle, sent its staff around the world finding timber from sustainably managed tropical forests. Such actions were successful in providing small quantities of 'sustainably' produced timber, and of raising the general profile of the issue, but also resulted in a string of opposing claims from established timber traders, as described above. As a result, ideas about timber certification were developed, and some schemes began to be developed.

Main principles for carrying out certification

The key to timber certification is the development of a system which combines two main functions: forest auditing, carried out in the forest or plantations of origin by an independent inspector; and timber tracing, to follow the movement of products from a certified forest. The combination of these two functions, if carried out successfully, would guarantee that a product comes from a well-managed forest. This will not necessarily be easy, and critics of certification have seized on the difficulties of providing proof, particularly in countries where well-established and widespread methods of illegal logging, processing and trading already exist, such as Thailand, the Philippines and Papua New Guinea. (Of course, illegalities are not confined to the South.)

In particular, new methods of tracing will be needed to ensure that illegally logged or uncertified timber is not added to shipments of certified timber during passage. A number of systems, including one for bar-coding logs, are under development. Some of them, developed by the Oxfordshire-based company Forest Log, have been used in trials to trace timber from tropical sources to retail shelves in the UK, and have proved successful. However, it is likely that some problems will remain and that a certain amount of fraud is to be expected within any certification system. While this may be an inevitable factor in any such system, the promoters of certification were convinced that any system had to be internationally credible.

THE FOREST STEWARDSHIP COUNCIL

The concept of an organization to evaluate, accredit and monitor wood and wood-product certifiers was first proposed by the Woodworkers' Alliance for Rainforest Protection (WARP), and in 1991 the idea of the Forest Stewardship Council (FSC) was developed. It was suggested that the FSC, an independent voluntary organization, would approve and *accredit* certifying agencies worldwide. The FSC intends to promote good forest management by evaluating and accrediting certifiers, by encouraging the development of national and regional forest management standards, and by strengthening national certification capacity by supporting the development of certification initiatives worldwide.

These agencies would, in turn, inspect and certify those timber producers which are operating their forests according to internationally agreed standards of good management. The model for organization is therefore:

Forest Stewardship Council
[providing accreditation for certification bodies]

↓

Certification bodies
[providing certification]

↓

Forest producers
[gaining certification and the right to label their products]

↓

Consumers
[gaining guarantees of good quality and ethically-produced timber][2]

The model for the FSC already existed, in the shape of certification schemes for organic or biological foods. A boom in consumer interest in organic food during the 1980s had resulted in a string of companies and NGOs running schemes for certifying that a farmer or grower really was producing food to agreed standards of organic agriculture. All the various certification schemes (of which there are now several dozen around the world) are affiliated to the International Federation of Organic Agriculture Movements (IFOAM), based in Germany. All certification schemes must have standards that satisfy the minimum laid down by IFOAM and the federation is charged with the task of policing the system. Recently a French certification company lost its licence after fraud was reported to the IFOAM secretariat. Although there is now a European Union-wide directive controlling trade in organic food, IFOAM's voluntary system provides the most widespread coverage of the global system.

The FSC was suggested as a way of setting up a similar system for timber. It soon became obvious that, if the scheme was to be worth undertaking, it would have to be dealing with a much larger market share than is the case for organic food. In the period building up to its launch, those interested in the FSC worked on two main issues: production of a working definition of 'sustainable forest management' – the Principles for Natural Forest Management; and the establishment of an international system for certifying 'well-managed forests' and tracing the timber from them. In addition, a number of basic management principles became clear during the preparation period:

❑ the FSC will only be effective if it receives support from all interested parties, including industry, NGOs, consumers and others, such as representatives of indigenous peoples' groups;

❑ setting standards will be a laborious and on-going process, and standards will continually have to be revised as new knowledge is acquired about management, ecology, etc;

❑ the FSC will also have to deal with governments and international agencies if it is to be successful;

❑ success requires credibility, and the FSC will need to be aligned with existing bodies such as the International Standards Organization (ISO);

❑ the FSC has to be a voluntary body, to avoid potential problems with GATT and the World Trade Organization;

❑ potential accreditation and certification schemes must be open to producers in tropical, temperate and boreal forests, to prevent market distortions developing; and

❑ the FSC must not be the property of a single organization or a single interest group[3].

The Founding Assembly

Following nearly three years of intensive preparation, the final steps towards creating a reliable timber labelling system were taken in Canada, in October 1993, when the FSC was officially inaugurated at its Founding Assembly. The Assembly consisted of 130 participants from 25 countries, including Argentina, Brazil, Colombia, Germany, Ghana, Honduras, Japan, Mexico, Papua New Guinea, Solomon Islands, Sweden, Switzerland, UK and USA. The participants, who represented a wide range of interests – governments, the timber trade, pulp and paper industry, environmental and other NGOs, community forestry groups, certifiers and indigenous people – voted unanimously to legally constitute the FSC as an independent, non-profit, non-governmental, membership organization. Some other elements of the Assembly were less successful and the meeting was characterized by some fairly deep splits, particularly within the NGO community. As this development is seen as a critical part of attempts to set standards of good forest management, the establishment of the FSC is reported here in more detail than has been the case for most other international initiatives covered in this book.

The Assembly voted to divide FSC membership into two chambers. The first consists of social, environmental and indigenous organizations and has 75 per cent of the voting power, while the second is drawn from individuals and organizations with an economic interest in the timber trade, and has 25 per cent of the voting power. Fears that the industry could eventually control the processes within the FSC led to the first major division at the Assembly. After prolonged arguments, a group of 13 individuals, including representatives of Greenpeace and FoE, decided to withdraw from the process and be considered as observers following the decision of the Assembly to allow representatives with economic interests to vote and serve on the board.

The Assembly also voted in a new board of directors consisting of nine individuals elected for a three-year term. Of these, two represent economic interests. These could include academics, consultants, representatives or employees of FSC certified companies, and representatives or employees of companies involved in the timber trade – producers, manufacturers, wholesalers, retailers – with a demonstrated commitment to the FSC Principles. There are rules for including both 'Southern' and 'Northern' economic representatives, while the remaining seven members come from environmental and social organizations and groups[4].

Forest Stewardship Principles

The Assembly ran into more problems when it came to agreeing the FSC principles and criteria for good management; ie the basic set of principles that any membership certification body would have to comply with in order to obtain endorsement from the FSC. The Assembly eventually voted to endorse the FSC Principles for Natural Forest Management, with the recognition that the draft needed further development. It would be fair to say that there were quite strong divisions about aspects of the principles and criteria – mainly regarding whether they were rigorous enough or not – and that while the FSC was indeed successfully launched and is being developed, there were some fairly serious birth pains. In spite of these divisions, all major international environmental NGOs now support the FSC and after 13 drafts, wide consultation and international review, the *Principles and Criteria* were approved by the founding members, in a postal ballot, by an overwhelming majority in June 1994. At the same time, the FSC's statutes, and *Guidelines for Certifiers* were also approved. The board of directors, with a mandate from the FSC membership, at this time began the process of legally establishing the FSC. The FSC has also begun the evaluation of certifiers for the purposes of accreditation and has begun to support many national certification initiatives worldwide, through workshops, training and technical support.

THE FSC MISSION STATEMENT

The Forest Stewardship Council shall promote environmentally appropriate, socially beneficial and economically viable management of the world's forests[*].
 In detail, this means:

❑ Environmentally appropriate management ensures that management practices maintain forests' biodiversity, productivity and ecological processes.
❑ Socially beneficial forest management helps both local communities and society at large to enjoy long term benefits from harvesting forests. Social benefits, such as employment, revenue, and guaranteed land rights provide strong incentives for local people to sustain the forest resource and adhere to long-term management plans.
❑ Economic viability means that forest operations are structured and managed so as to be sufficiently profitable, which enables stability of operations and genuine commitment to principles of good forest management. Economic viability does not mean economic profit at the expense of the forest resource, the ecosystem or affected communities. While recognizing the limitations of the marketplace, the tensions between the necessity of adequate financial returns and the principles of environmentally and socially responsible forestry operations can be reduced through efforts to market products for their highest and best value, and incentives (such as certification) for capital reinvestment in the forest resource[**].

Sources: [*] FSC (1995) *Fact Sheet*, Oaxaca, Mexico; [**] FSC (1994) *Fact Sheet*, Richmond, Vermont

---

The FSC Principles

The FSC Principles and Criteria are not a set of standards as such, but define the *minimum* requirements for such standards. Certification bodies are free to set more stringent standards if they or their customers so desire, but to gain

FSC accreditation they cannot fall below those set by the FSC. There are nine Principles for Forest Management agreed and a draft tenth relating to plantations:

❑ **Principle 1: Compliance with FSC Principles** Forest management operations shall respect all applicable laws of the country in which they occur and international treaties and agreements to which the country is a signatory, and comply with all FSC Principles and Criteria.
❑ **Principle 2: Tenure and Use Rights and Responsibilities** Long-term tenure and use rights to the land and forest resources shall be clearly defined, documented and legally established.
❑ **Principle 3: Indigenous Peoples' Rights** The legal and/or customary rights of indigenous people to own, use, and manage their lands, territories and resources shall be recognized and respected.
❑ **Principle 4: Community Relations and Workers' Rights** Forest management operations shall maintain or enhance the long-term social and economic well-being of forest workers and local communities.
❑ **Principle 5: Benefits from the Forest** Forest management shall encourage the optimal and efficient use of the forest's multiple products and services, to ensure economic viability and a wide range of environmental, social and economic benefits.
❑ **Principle 6: Environmental Impact** Forest management shall conserve biological diversity and its associated values, water resources, soils, and unique and fragile ecosystems and landscapes, and, by so doing, maintain the ecological functions and integrity of the forest.
❑ **Principle 7: Management Plan** A management plan – appropriate to the scale and intensity of the operations – shall be written, implemented and kept up to date. The long-term objectives of management and the means of achieving them, shall be clearly stated.
❑ **Principle 8: Monitoring and Assessment** Monitoring should be conducted – appropriate to the scale and intensity of forest management – to assess the condition of the forest, yields of forest products, chain of custody, management activities and their social and environmental impacts.
❑ **Principle 9: Maintenance of the Natural Forest** Primary forest, well developed secondary forest, and sites of major environmental, social or cultural significance shall be conserved. Such areas shall not be replaced by tree plantations and other land uses[5].
❑ **Principle 10: Plantations** Plantations should be planned and managed in accordance with Principles 1–9 and the following criteria: such plantations can and should complement natural forests and the surrounding ecosystem; provide community benefits; and contribute to the world's demands for forest products. (Draft principle for plantations not yet ratified by FSC membership.)

In addition to these principles the FSC has also developed a set of guidelines for certifiers seeking accreditation under the FSC system.

How the FSC will work
The FSC is based in Oaxaca, Mexico. Producers whose forest management systems have been FSC certified, will be permitted to label their products with an internationally recognized symbol or logo. This will allow consumers to buy products with confidence, knowing that the label is reliable and that the product really does come from a well-managed forest. Certifiers of forest products will be evaluated by the FSC on the basis of: their adherence to the FSC Principles; their adherence to FSC guidelines for certifiers; and their *specific operational standards* of forest management, which are locally-appropriate standards and must be approved by the FSC.

INDEPENDENT CERTIFICATION BODIES

The FSC will oversee the process of timber certification and product labelling worldwide. However, the field inspections and the actual certification will be carried out by a range of independent certification bodies, affiliated to and approved by the FSC. Ideally, in time most major timber producing countries will have at least one, and probably several, certification bodies. Membership status is open to a wide range of organizations and individuals, representing social, economic and environmental interests.

There are already several organizations which are certifying areas of forest. These have agreed to meet and conform to the FSC principles and criteria. Until the formation of the FSC, these organizations may well have been using different standards for auditing forests and different systems for tracing timber. Organizations include:

❑ the Smartwood Program of the Rainforest Alliance in the USA, which has grown out of a tropical forest campaign group;
❑ the Soil Association's Woodmark system in the UK, drawing on the long-standing expertise of an existing organic certification body[6];
❑ SGS Forestry (formerly Sylviconsult) in Oxford, UK, an affiliated company of SGS International based in Geneva, which has 1170 offices worldwide, 309 laboratories and 32,000 staff certifying a wide range of products.

During 1993, a series of FSC consultations were carried out in a range of countries, including the UK, Switzerland, Sweden, Ghana, Malaysia, Brazil, Peru, the Pacific Northwest of the USA and Canada, to ascertain interest in certification and to canvass opinions on the different systems proposed. In addition, FSC programmes are being devised, through NGOs, industry and government cooperation, in Finland, Indonesia, Sweden and Switzerland.

Reactions to the FSC have been mixed. It would be fair to say that initial reactions from many people in the timber trade, and in relevant government departments, were negative. However, this situation appears to be changing fast, as the system gathers momentum and members of the international timber trade become fearful of being left behind or swept aside by other, certified, sources. Similarly, government departments, which had at first proved reluctant and somewhat resentful of what they regarded as interlopers in their traditional areas of expertise, have in several cases since taken an active interest in development of certification systems. For example, some governments, including that of Austria, have openly supported and help finance the FSC's development.

WWF has recently stated:

In the absence of a legally-binding international forest agreement, we believe that industry has a responsibility to ensure that the timber it uses comes from well-managed forests – for two reasons. Firstly, any company which ignores the sustainability of supply of its raw material base is very vulnerable, especially if, as with forests, the resource is being over-exploited. Secondly companies have a responsibility to carry out their business in a way which reduces their impact on the environment. Shareholders and customers are increasingly demanding this.

The FSC will take the guesswork out of deciding whether a particular shipment of timber has come from a well-managed forest. We urge companies to investigate how they can use the FSC and the certifying organizations to help them ensure their timber purchases only come from well-managed forests. WWF has been working with a number of companies which have been trying to do exactly that.

In conclusion, WWF encourages companies, City and County Councils, architects and specifiers to adopt purchasing policies which clearly state that they will phase out the specification and use of all timber products which do not come from an independently-certified, well-managed forest by 31st December 1995. This will give those forest producers who have been certified as managing their forests well and those manufacturers who buy from these forests, a market advantage.

More specifically WWF calls on companies, councils and architects and specifiers to adopt a timber product purchasing policy which:

❏ *gives preference to suppliers who can provide independent evidence of good forest management*
❏ *gives preference to 1995 Group members*
❏ *ignores timber labels and unverified certificates of sustainability*

and once the Forest Stewardship Council is fully operational:

❏ *requires that all wood product sources are certified by an independent body which has been accredited by the Forest Stewardship Council*[7].

By January 1995, more than 4 million ha of forests had been independently certified from 17 countries worldwide, including ten in the South and
seven in the North. Already numbers of retailers in the UK and the USA are
selling wood and wood products from these forests.

The future of the FSC remains volatile, and some sections of the industry
would clearly like to see it fail. In June 1995, a joint Canadian-Australian
proposal was made to a meeting of the International Standards Organization
in Oslo, suggesting the development of an Environmental Management
System (EMS) standard. The proposal, originating with the Canadian
Standards Association, would have been in conflict with the FSC, and was
suggesting standards that were considerably less stringent than the FSC's own
Principles and Criteria. In the event, sustained lobbying from WWF and other
organizations resulted in the proposal being withdrawn[8].

THE WWF 1995 GROUP

The FSC will work effectively only if it has support from industry. In the
UK, a partnership with forward-looking sectors of the UK timber trade has
been formed. Since 1991, WWF has been working closely with a number of
companies which have committed to phasing out, by 31 December 1995, the
sale and use of all wood and products which do not come from independently certified well-managed forests. This has become known as the *WWF 1995
Group*. By July 1995, 5 companies had joined the group, from retailing giants
such as Boots the Chemists, through major purchasers such as British Rail, to
small wood-using companies like David Craig.

Requirements for joining the WWF 1995 Group, as revised at a mini-
seminar in London on 17 January 1995, include:

1   Commitment to the FSC as the only currently credible independent certi-
    fication and labelling system.
2   Commitment to the phasing out of the purchase of wood and wood
    products which do not come from well-managed forests as verified by
    independent certifiers accredited by the FSC.
3   The phasing out of the purchase of wood and wood products that do not
    come from well-managed forests and the phasing in of wood and wood
    products which can be shown to be from well-managed forests by 31
    December 1995. In practice this means:
    (a) A proportion of wood and wood products will be certified as coming
        from well-managed forests as defined by the FSC, by independent cer-
        tifiers accredited by the FSC. The proportion of wood in this category
        should be demonstrably increasing.
    (b) Remaining wood and wood products will come from known forests
        which the Group member has demonstrated are 'well-managed'. The
        proportion of wood in this category should be demonstrably increasing.

(c) Wood and wood products which cannot be traced to known forests and/or where the quality of management is in doubt will be eliminated.

4 A named senior manager will have responsibility for implementing the above commitment. Progress towards the target will be monitored via six-monthly progress reports.

5 WWF 1995 Group members may use the FSC logo when they are licensed to do so. Other labels denoting well-managed sources will not be used[9].

The impact of the WWF 1995 Group is already being widely felt. Eight of the WWF 1995 Group members – Boots the Chemists, B & Q, Do It All, Great Mills, Homebase, MFI, Texas Homecare and Wickes Building Supplies – have signed a position statement setting out their agreed policy on wood procurement. They have all agreed not to buy products from unknown sources or poorly managed forests beyond the end of 1995[10]. The WWF 1995 Group trades over £1 billion worth of wood and wood products every year, which is approaching 10 per cent of total wood consumption in the UK. More than 35 million customers a week shop in those stores.

The WWF 1995 Group will continue to exist after the target date, and will then include European and international members. Membership of the group after the target date will be dependent on their having already reached the target. In return, WWF provides technical support to members, information, including marketing information, and some promotion.

The companies have drawn up timber product purchasing policies which clearly state that unknown or unsustainable sources of wood and wood products will be phased out by 31 December 1995. Membership of the group requires active commitment. A number of companies which initially joined, but failed to meet the requirements for membership, have since been delisted. Group members have been active in seeking certified, or certifiable, sources of timber and for example took part in a field trip to Finland in May 1994 to discuss options for certification with the Finnish timber trade[11].

Support for the 1995 Group has been increasingly vocal among the membership. B & Q has become a major force behind the establishment of the FSC. Martin Laing, chairman of John Laing plc, wrote in March 1994:

I implore all wood using companies to ... be bold and join WWF's 1995 Group, which is committed to phasing out the use and sale of wood and wood products that do not come from well-managed forest by 31 December 1995 – as Laing Homes already has. We should see environmental achievement as a welcome business opportunity – not a threat[12].

Members of the 1995 Group: July 1995
Acrimo Ltd; Akzo Nobel Decorative Coatings UK and Eire; B & Q plc; BBC
Magazines; Bernstein Group plc; Bioregional Charcoal Co Ltd ; Boots the
Chemists; Borden Decorative Products Ltd; British Rail; Richard Burbidge
Ltd; Chindwell Co Ltd; Core Products Ltd; David Craig; Crosby Sarek Ltd;
Do It All Limited; Douglas Kane Hardware; Dudley Stationery Ltd;
Ecological Trading Company; Forbo Lancaster Ltd; Richard Graefe Ltd;
Graham and Brown Ltd; Great Mills (Retail) Ltd; Green Life Marketing;
Harrison Drape; Helix Lighting Ltd; Homebase; Indian Ocean Trading Co;
John Dickinson Stationery Ltd; Larch-Lap Limited; Laing Homes Ltd;
Magnet Ltd; F W Mason & Sons Limited; MFI; Milland Fine Timber; M & N
Norman; Moores of Stalham (UK) Ltd; Premium Timber Products Ltd;
Rectella International Ltd; Rothley Limited; J Sainsbury plc; FR Shadbolt &
Sons Ltd; Sherwood Promark; Shireclose Housewares Ltd; WH Smith
Business Supplies; WH Smith Retail; Spur Shelving ; Swish Products Ltd;
Tesco plc; Texas Homecare Limited; Vymura plc; Wickes Building Supplies
Ltd; John Wilman Ltd; Woodbridge Timber Ltd; Woodlam Products.

CONCLUSIONS

Market-led solutions are, as stressed before, not a universal solution.
Nonetheless, the positive responses from both industry and consumers lead us
to believe that certification is an important part of the process of reform.
Considerable progress has been made since the early 1990s when many
people were questioning the validity of the concept of certification. Today the
question is how certification will be implemented rather than if it will exist at
all. As such, it must take its place alongside policy reforms, voluntary reforms
by governments and industry, and other independent systems such as those
being developed in Indonesia. Some first thoughts towards the make-up of an
overall strategy are contained in Chapter 10.

10
A Strategy for Forests

The preceding text has argued that there are a series of real problems regarding the social and environmental effects of the timber trade, but also that there are some potential solutions to these difficulties. These solutions are neither necessarily utopian nor exclusive, and could potentially result in a 'win-win' situation with respect to both forests and industry. This chapter offers some solutions, both in general terms of what is needed and specific action points for named interest groups.

The suggestions below draw heavily on a *Global Forest Strategy* developed by WWF International[1] and also on the collective efforts of the UK Forest Network[2] and from an investigation of timber labelling carried out by Mike Read for WWF UK[3].

A VISION OF FORESTS IN THE FUTURE

This book began with a vision of the timber trade, and of ways in which it could continue providing resources, jobs and a financial return without ruining the world's forests. However, as described in Chapter 2, the timber trade is only one part of a complex series of pressures on the global forest estate. Any overall strategy for addressing problems in the world's forests needs to be drawn up within a framework of ideas about how the forest estate *might* look in the future. Over the past two or three years WWF International, in cooperation with its 28 national organizations and numerous programme offices, has drawn up the following vision for forests in the future.

WWF'S FOREST VISION

In view of the increasing rates of deforestation and the loss of quality in tropical, temperate and boreal forests, WWF believes that it is essential to maintain, and where necessary restore, the global forest estate, such that it meets a wide potential range of human and non-human needs. Forest management systems must be based on the principle of sustainability, ie they must be environmentally appropriate, socially beneficial and economically viable. A pre-requisite for maintaining the

> *multiple value of forests is the conservation of biodiversity at the genetic, species and ecosystem level. This will require a 'paradigm shift' in the thinking of foresters, who still tend to give priority to timber production and seek to maintain the flow of other goods and services only when they are compatible with this. In future, primacy should be given to the conservation of biodiversity and forest functions as these are the basis upon which any human use of forests depends.*

This emphasis is distinct from that followed in most countries today, where forests are seen primarily as resources to be mined or farmed for timber and pulp. Other human and non-human goods and services, such as environmental protection, recreational facilities, non-timber products, etc are therefore regarded as secondary to these functions.

It should be noted that no fully-developed blueprint for sustainable forest management currently exists. Modifications will evolve as part of a process of developing new management techniques, including systems of co-management, although some must be implemented with urgency if important forest functions are not to be lost. Although much policy development remains to be carried out, we can start to draw a general picture of the global forest estate that WWF hopes to see in the future.

> *In future, forests will, on a global scale, be both more extensive and of a higher quality than at present. They will contain a greater proportion of natural forest, including the majority of existing old-growth forests, augmented by restored secondary forest. Ecologically and socially appropriate reforestation (using mixes of native species), coupled with sustainable forest management will be widespread.*
>
> *A proportion of the world's forests is likely to remain under fairly intensive timber production for the foreseeable future, but this will be based on more rigorous social and environmental safeguards and the principles of landscape ecology. Independent certification will play an important role in encouraging improved management.*
>
> *A range of management approaches involving the full participation of local communities will be developed, based on the concept that local people must be the prime beneficiaries of forest management. Non-timber values will be given greater priority and the impacts of pollution will be reduced. At the policy level, regional and cross-sectoral collaboration will be enhanced to conserve forests.*

It is clear that achieving this vision will be possible only if other, wider issues are also dealt with, including such problems as inequality in access to land and resources; repression of women, minorities and indigenous people; concentration of political and economic power; and world consumption patterns. However, in this chapter a fairly narrow, forest-based, focus is maintained, and elements of WWF's forest strategy that have direct implications for the timber trade are outlined.

WWF'S GLOBAL FOREST STRATEGY

The following points are not listed in order of priority.

Objective 1: Establishment of a network of ecologically-representative protected areas

This objective addresses the issue of protected forest areas for the conservation of *biodiversity* and the maintenance of *ecological processes*. The focus is consequently on preserving sufficient forest to allow natural ecological dynamics to continue indefinitely. This means that the area of protected forest must be determined on a case-by-case basis, rather than for example by aiming to set aside a standard area of forest per country. Protected forests will need to be securely funded as a step towards guaranteeing their long-term security. Many protected forests will have people living in or around them, including indigenous people. Local involvement, in both planning and co-management, is therefore essential to success, ie in many cases protected forests will not be strictly and exclusively nature reserves (although reserve areas are undoubtedly needed within this system), but will be protected from large-scale or commercial exploitation.

This network will provide a basic reservoir for forest biodiversity and natural processes, and also a reference against which other, more managed, forest systems can be compared. Developing such a network, from a current situation where many protected area systems are developed on a slightly *ad hoc* basis, will involve identification of forest types and a review of their original, current and likely future extent, including support for reliable forest inventories, geographical information system (GIS) mapping, gap analysis, etc. Estimation of the area of each forest type requiring total protection to ensure the preservation of biodiversity and forest dynamics will also be needed, and priorities for new protection set.

Objective 2: Achievement of environmentally appropriate, socially beneficial and economically viable forest management outside protected areas

The aim of this objective is to improve *forest quality* in forests outside areas fully protected for biodiversity* including restoration of forest quality in existing stands and landscapes where necessary. It recognizes that biodiversity and environmental protection objectives cannot, in most countries, be met solely with a protected-area network and there is consequently a need to reduce the contrast between reserves and managed areas. Forests managed for other

---

* WWF has defined a series of criteria of forest quality, based on general principles of *authenticity, forest health, environmental services* and *other social and economic benefits*, and these will be further developed to take into account methods of quantification and measurement. Reference N Dudley, J-P Jeanrenaud and S Stolton (1993) *Towards a Definition of Forest Quality: Proceedings of a Colloquium*, WWF UK, Godalming, Surrey.

services must, therefore, also include environmental considerations as important components of any management plan.

Forests should be managed for a variety of values. All the human needs that forests can supply, including resource provision, social, economic, aesthetic, recreational and spiritual goods and services, should be included when considering forest management. A proportion of forests should be managed primarily for watershed protection, soil conservation, carbon storage, etc. This is *in addition* to forests set aside for biodiversity protection under Objective 1. Indeed, a large proportion of the global forest estate is simply unsuitable for commercial timber management because of its location and climate, and the other values that it provides.

Any management that does take place should meet FSC Principles and Criteria or the equivalent. It should emphasize a landscape ecology approach, avoid forest fragmentation and take account of issues such as biodiversity, environmental services and the likely effects of global change.

Management for commercial purposes should now be aimed, wherever possible, at secondary forests and only used in natural forests after careful assessment of local needs. Management must also ensure that forest dwellers do not suffer physically or emotionally as a result of mismanagement of, or bad planning decisions relating to, forests.

Objective 3: Development and implementation of ecologically and socially appropriate forest restoration programmes

This objective aims to *re-create* original forests where these have been degraded or destroyed, and to *enlarge* total national forest area where this has fallen below optimal levels. Ideally, forest restoration should entail re-creation of something as near to the original forest as possible. In some cases this will be essential, for example when insufficient natural forest remains to meet Objective 1. In many cases, forest restoration can also include a commercial return, but this should not be the sole function of restored forests.

Forests should be restored under criteria and standards of high forest quality, stressing multiple use and restoration of natural dynamics and biodiversity. Use of local and native tree species should be encouraged, and a ban on planting exotics may be appropriate in some areas. Forest restoration programmes should use local provenances and avoid creating large genetically identical stands. Human needs, particularly those of indigenous and local people, should be included in all forest restoration plans, through a process of consultation, joint decision-making, co-management and benefit-sharing. Plantations should not be created through replacement of natural forests or other important ecosystems. The possibility of restoration should not be used as an excuse for further deforestation.

Objective 4: Reduction of forest damage from global change

This objective aims to reduce and where possible eliminate the impact of a range of global change effects, including pollution, ozone depletion, global warming, human-induced forest fires, etc. It therefore relates less directly to the timber trade as a *causal factor*, although the forestry industry is currently losing productivity in some areas due to the impact of a range of air pollutants. (This is sometimes partially offset by fertilization from pollutants such as nitrogen oxides.)

A major factor is elimination of pollution damage from local and long-range atmospheric pollution, water pollution, soil contamination and toxic waste. One way to measure success in this area is by using the theory of *critical loads*; these are defined as the amount of pollutant that can enter an ecosystem without disrupting natural processes. It should be noted that critical loads still have not been established for many pollutants and ecosystems. Reducing pollution damage will generally involve reducing pollution at source, although in some cases remedial action within forests will be necessary. Planning for global warming is essential in forestry, but it will be some decades before we can hope to measure the success of any efforts.

Objective 5: Use of forest goods and services at levels that do not damage the environment, including elimination of wasteful consumption

This objective aims to establish levels of consumption of forest products that minimize environmental effects, and to attain a level of use of forest goods and services within the regenerative capacity of the forest estate, in the context of the overall forest strategy. This must be done bearing several other factors in mind. These include the effects of over-consumption of forest products, which might argue for a *reduction* in overall use in some cases, but also the potential for forest products to *substitute* for other products that may have even greater environmental effects. Within the current context of the timber industry, this might for example mean arguments for less use of paper, but greater use of high quality timber, to substitute, for example, for aluminium window frames, other building materials, plastic and brick.

Building on the existing targets, but with the aim of providing greater focus, two new campaign objectives have now been set. These will help focus the activities of the campaigning work on forests over the next five years:

❑ Establish an ecologically representative network of protected areas covering at least 10 per cent of the world's forests by the year 2000, demonstrating a range of socially and environmentally appropriate models.
❑ Ensure the independent certification of 10 million ha of sustainably managed forest by 1998.

ACTION PLAN FOR IMPLEMENTING THE FOREST STRATEGY

The strategy outlined above will simply remain a list of 'coulds' and 'shoulds' unless clear action points are identified, prioritized, allocated to specific institutions, and lobbying and pressure applied to ensure that they are carried out in practice. The task of restoring a healthy, sustainable and high quality forest estate is beyond the capabilities of any one organization or country, and will require willingness and effort on the part of many different interest groups. In the following section, an attempt is made to identify actions from international bodies, national governments, industry and NGOs.

Unfortunately, past attempts to address environmental problems have, all too frequently, been imposed in so heavy handed a manner that they alienate a large proportion of other interests groups. A national park created by expelling indigenous people is not an adequate solution, any more than forest management programmes that take no account of local needs and desires, or controls so inflexible that industry cannot operate. Any solution will inevitably be some kind of compromise, and the debate between industry and environmentalists will most likely be ongoing.

Any action to address forest problems must, therefore, take into account a number of important background *principles*. These can act as a series of *aides-mémoire* both to ensure that environmental problems are not addressed at the expense of other legitimate interests and to make certain that all the necessary environmental issues are considered. Such principles might include: equity in access to resources; use of principles of fair trade; introduction of a system for real valuation of forest products; a long-term plan for a sustainable system for any forest operation; ensuring the establishment of a permanent forest estate; full local consultation and respect for indigenous and local rights; and provision of adequate protected areas. (For a more complete overview, see the FSC's Principles and Criteria, Chapter 9.)

MAIN PARTICIPANTS INVOLVED IN SETTING THE AGENDA

The timber trade is controlled and/or influenced by a wide range of different international, national and local interest groups; some of the main ones are:

❑   **International** bodies such as:
   –   international treaty and secretariat organizations including the Convention on Biological Diversity, CSD, the Helsinki Process, etc;
   –   United Nations bodies such as the FAO and UNESCO;
   –   multilateral development banks such as the World Bank, Inter-

American Development Bank, the European Bank for Reconstruction and Development (EBRD), etc;

–   bilateral aid organizations such as the UK Overseas Development Agency, US Agency for International Development (USAID), the Swedish International Development Authority (SIDA), Finnish International Development Agency (FINNIDA), etc; and

–   commodity agreements such as the International Tropical Timber Agreement (ITTA).

❑ **National** bodies, mainly governments and state forestry agencies.

❑ **Trade associations**, including national and international bodies and some specific 'campaign-orientated' organizations, such as the UK TTF and its *Forests Forever* campaign and the Finnish Forestry Association's *Plusforest* campaign.

❑ **The timber trade**, including forest companies, wood processing companies, specific procurement organizations, manufacturers using timber and pulp and retailing outlets.

❑ **Non-governmental organizations** (NGOs), including international bodies such as WWF, Greenpeace, FoE, WRM, World Resources Institute etc, and many regional and national NGOs.

❑ **Consumers**, represented to some extent by consumer organizations, many of which belong to the International Organization of Consumer Unions.

SUGGESTIONS FOR ACTION

The following list of suggestions provides a draft action plan for some of the key interest groups outlined above.

International organizations

Commission on Sustainable Development
The CSD is the clearest and most comprehensive of a series of policy initiatives regarding forests. However, its remit and intentions remain unclear. The CSD therefore requires a clear statement of intent. We suggest that its aim should be *to maintain, and where necessary expand, the global forest estate, and to ensure that forests are managed sustainably for the complete range of potential goods and services.* Part of the progress of setting the agenda for the CSD should be to help determine the status and future role of other forest initiatives, to reduce the confusion and duplication that is currently occurring.

We believe that the CSD must urge governments to commit themselves to clear and quantifiable targets. Despite the myriad of new forest initiatives that have emerged since the Earth Summit (UNCED) in Rio de Janeiro, 1992, there has been a reluctance to set precise targets and governments have preferred vague statements of intent. Without specific goals it is impossible to measure progress systematically. WWF therefore proposes that governments implement the following targets by the year 2000:

❑ establish a network of adequately protected, ecologically representative forest areas – at a minimum 10 per cent of original forest cover;
❑ ensure that forest management outside protected areas is environmentally appropriate, socially beneficial and economically viable.

The CSD should play an important role in promoting these targets by mobilizing adequate funding and monitoring progress.

We believe the international cooperation in forest issues is vital. A number of international agreements already exist which if implemented would greatly assist efforts to improve forest conservation and sustainable management. A new legally binding instrument is therefore not a priority, but, *the full implementation of existing agreements is crucial*.

NGOs have an ambivalent attitude to the CSD, following the disappointment regarding forests arising from UNCED and a generally lacklustre approach to the issue up to now. Support in the future will come only if the CSD adopts an approach similar to that outlined above.

Other international initiatives
Several other international conventions have an actual or potential impact on forests. The **Convention on Biological Diversity** and the **Climate Change Convention** both have the option of developing forest protocols. The former already has a role in protecting critical forest habitats in some regions. There is still a considerable move towards developing a Global Forest Convention although there is no agreement about form or content as yet. All of these have the possibility of delivering useful benefits for forests, but also of achieving little, or perhaps causing damage, if set up or applied in the wrong way. Therefore, we suggest that support for these initiatives should be dependent on their supporting the general principles outlined in this chapter, and summarized in the CSD section above.

The **Convention on International Trade in Endangered Species** (CITES) is an important tool for monitoring the trade, and controlling this when it is causing net disbenefits. To carry out this role effectively, the CITES secretariat and representatives need resources; development of more effective timber tracing is one of the keys to effectiveness in this area. More timber species, including temperate timber species, should be listed in CITES, and CITES should be implemented rigorously for timber species.

The **Tropical Forests Action Programme**, launched with much fanfare a few years ago but gradually abandoned by many of the sponsoring organizations, should be wound down. It has now been superseded by FAO's role in the CSD process, and existing resources should be transferred to this activity.

United Nations bodies

The **Food and Agriculture Organization** (FAO) is the leading body in the CSD. The FAO's role in forestry over the past few decades has done little to develop confidence that it will be able to fulfil this role adequately. Considerable and continuing changes, with respect to both attitudes and operational practice, will be needed if it is to meet the challenges outlined above.

Other United Nations bodies with a role in developing a truly sustainable forest estate include the **UN Environment Programme** (UNEP) and the **UN Commission on Transnational Corporations** (UNCTC). UNEP should be chiefly involved through its field programme and monitoring activities. The UNCTC should further develop the Code of Practice for transnational companies, originally drawn up for UNCED, which would apply to the activities of the international timber trade.

Aid organizations

There is potential for **multilateral development banks** (MDBs) to take a positive role in forest development. However, past experience has shown that changes in practice will be needed if this is to be possible, and if negative impacts are to be avoided.

MDBs such as the various arms of the World Bank, the African, Asian, and Inter-American banks, the EBRD, etc, should develop clearer policies to avoid funding unsustainable or damaging projects. Environmental assessments, carried out well before funding is approved, should include full local consultation, transparency and accountability, and projects must include elements of local management and control. Information relating to board decisions of MDBs should be made public. MDB forest policies should apply to non-forest sectors to ensure that lending in other sectors does not promote deforestation, directly or indirectly. Environmental costs of projects should be internalized. All MDBs should have a policy not to fund logging in primary and old-growth forests.

Similar considerations apply to other multilateral and bilateral aid agencies, such as the European Development Fund and individual national programmes including the Danish Aid Agency (DANIDA), FINNIDA, the Norwegian Aid Agency (NORAID), the UK Overseas Development Agency (ODA), SIDA, USAID, etc. Aid disbursements should not be tied to trade or other advantages for industry in the donor country. The emphasis of funding should shift from individual projects to longer-term areas such as capacity and institution building, training and public education.

Commodity agreements

The **International Tropical Timber Agreement** (ITTA) should be widened to include all timbers, an aim that remains important despite the failure to decide this at the last renegotiation in 1994. The ITTA, and the associated ITTO, should also be relieved of any environmental mandate or, if such a directive does remain, this should be implemented and enforced.

National organizations

Governments

Despite the growing internationalization of decision-making, national governments play a key role in setting policies towards forests within their own countries and in controlling trade. Within the particular context of the timber trade, several points are important.

Governments set the policy framework within which the timber industry operates. Current developments within the CSD, and the other initiatives developing criteria for sustainable forest use, will help provide the necessary information for devising within-country policies for high quality forests. Such development necessarily comes from government, although also involves input from the timber industry, research and academic bodies, consumers organizations, regional government and NGOs.

Governments can also help control and monitor trade from abroad. The import of timber should only be allowed if it can be proved to come from legal sources. For example, the import of all tree species listed as *endangered* under the IUCN World Conservation Union definition should be prohibited, unless they come from certified sources or well-managed plantations. Governments can respect and support existing producer-country legislation by applying reciprocal import bans on species and timber products banned for export for conservation reasons by individual producer countries. Governments can also help balance trade and conservation interests by insisting that companies based in their country fulfil that country's environmental and social regulations, *wherever* in the world they operate.

Within the country, action is needed to control misleading claims made about the environmental-friendliness of timber and paper products. This could be achieved through a strengthening of such institutions as the UK Advertising Standards Authority and Trade Descriptions Act. Individual governments can help develop timber certification by working with NGOs and the timber trade, in cooperation with the FSC.

Lastly, sustainable utilization of timber products could be actively encouraged, beginning with a reduction in consumption, especially of short-term timber uses such as disposable goods and advertising products, and introduction of incentives for businesses and local authorities to start recycling schemes. If correctly produced, timber could be promoted as a high quality product, made to last. Schemes for recycling paper, timber and other products often require further encouragement, where necessary backed by legislation.

Trade associations

Timber trade associations should be leading the industry, taking an innovatory role and representing the interests of the trade from a conservation standpoint. Currently, they frequently make little attempt to provide leadership for their members, particularly with respect to environmental issues, but instead follow behind their more innovative members. Useful functions for trade associations might include support for initiatives such as the FSC, encouraging an open dialogue with NGOs, and ensuring that the trade does not undermine the resources that, ultimately, it depends upon.

A more substantial, respected, trade association could lift the whole level of a country's timber industry. Membership would be dependent on a proven track record of trade carried out in an environmentally and socially responsible manner – thus reflecting best practice – and only those operations attaining high enough standards would be admitted. The trade association might thus act as an incentive, rather than a constraint, on progress. The alternative, where trade associations promote poor forestry practice and in some cases actively attack environmental groups, will hamper developments towards a healthy global timber market.

Trade

If the changes suggested above took place, the timber trade would be operating in a very different national and international climate. Those parts of the trade which had developed environmentally and socially sustainable operations would be at a clear advantage, while those operating recklessly or illegally would be under considerable pressure. Unfortunately, this is still often not the case today.

Elements of the timber trade can, and already are, developing along lines that we have suggested are necessary to secure a long-term, high quality global forest estate. Development of clear policies, for harvesting, forestry operations, purchasing, etc are important steps along the way to providing a substantial improvement in policy. Removal of misleading or unsubstantiated claims from timber products is an early prerequisite of progress.

We suggest that, ultimately, the only way in which this can be systematically applied and verified is through a system of certification, organised by a national certification programme under the auspices of the FSC.

Non-governmental organizations

The large range of NGOs involved in the forest debate allows different organizations to fulfil different functions in terms of lobbying, education, etc. It also, importantly, allows NGOs to adopt different positions with respect to the radicalism of their demands and their mode of action. However, this varied approach will only remain effective as long as there is effective coordination between different NGOs.

NGOs fulfil the function of lobbyists and educators within general efforts to improve forest policy. They will continue to expose and campaign against environmentally and socially inappropriate forest operations, and to promote good practice in national and international initiatives. Specific NGOs can also help by promoting joint initiatives with industry where appropriate, and by providing support for the FSC and for traders dealing in timber and wood products that have been certified by an FSC-accredited certifier. As part of this action, NGOs and individuals can help stamp out misleading claims by making full use of national and international legislation regarding advertising claims. The FSC is an NGO with a particular role to play with respect to the timber trade and we regard its further development as critical to the success of timber certification.

11
Conclusions

For the past fifteen years, the timber trade has been remarkably successful in convincing governments and the public that it plays a negligible role in forest loss. Representatives have argued thaat most deforestation is caused by agricultural clearance or fuelwood collection. Population growth, rather than industrial exploitation, has been blamed as the underlying problem. In this analysis, it is claimed that the impact of commercial logging only accounts for a few per cent of global issues.

Successive independent investigations, including earlier work by the authors, have cast doubt on this claim. Research for the current book has gone further, and leads us to a conclusion that directly contradicts that of the timber trade. If we consider the world's most ecologically important forests, and take the survival of biodiversity as a major criterion, our assessment suggests that in fact *the timber trade is currently the most important cause of forest degradation around the world*.

Our very different conclusion is based on a number of important considerations, which have generally been overlooked in assessments made by the timber trade. These include:

❑ comparing the location of timber trade operations with areas of high biological wealth, particularly in primary forests;
❑ looking at forest *quality* as well as *quantity*;
❑ extending the assessment to *all* forests, rather than just tropical moist forests;
❑ including an assessment of illegal logging;
❑ incorporating information about more general changes in global forest conditions.

Each of these factors is examined in more detail below.

QUALITY AND QUANTITY

When a previous WWF report on temperate and boreal forest was published in 1992, one of the most contentious sections dealt with events in British Columbia. Indeed, the day after publication three members of the 'BC Forest Alliance' flew to London for what proved to be a fairly stormy meeting at the Canadian High Commission, with representatives from WWF, Greenpeace and the Women's Environmental Network. A major point of contention was our reference to literature which called Canada the 'Brazil of the North' and compared British Columbian clearcuts with Amazonian deforestation. The industry argued that the comparison was invalid, because in Canada the forest was either replaced or regenerated naturally, whereas in Brazil it tended to be converted to ranchland or farms.

This misses a fundamental point. From the perspectives of biodiversity and environmental balance, there is often little to choose between felling a forest and either replacing it with a tree plantation or converting it to farmland. In either case, the vast majority of the original native wildlife species do not survive. Even if total *number* of species remains constant after felling (and as described earlier, in some cases it even apparently increases), the rare natural species are likely to be replaced by aliens, weed species and primary colonizers. Overall biodiversity is reduced at a landscape level. Loss of forest quality has already occurred in large parts of Europe, North America and Australasia. It is becoming increasingly significant in several Southern countries as well, such as Kenya, Rwanda, Malaysia, Indonesia and Chile.

Analysis of the timber trade's impact should not, therefore, simply look at loss of area under trees, ie only at total deforestation. It also has to consider the *biological quality* of the forest that remains.

INCLUDING ALL FORESTS IN ASSESSMENTS

The major emphasis put on tropical rainforest loss during the 1980s meant that problems in many other forest areas were overlooked or underestimated. It also meant that representatives of the timber trade, and other industrial sectors, could focus their efforts on avoiding criticism of their operations in tropical moist forest and ignore events in other areas. Some companies even went to great lengths to refine and sometimes redefine 'tropical moist forest' so that its theoretical borders fell outside their own operational areas. Others claimed 'we don't sell tropical timber' as if it were a major environmental recommendation, without saying what they *did* sell, which might have been old-growth temperate or boreal timber. The implication seemed to be that any damage done to sub-tropical forests or tropical dry forests was of comparatively little importance.

Our analysis, on the other hand, has looked at *all* forests: tropical moist forests, tropical dry forests, open woodland, sub-tropical, Mediterranean, temperate and boreal forests. Once these are included, the role of the timber trade immediately grows in significance. Unlike tropical moist forest, where there have been endless arguments about cause and effect in forest loss, the chief cause of loss of natural and semi-natural forest and degradation in temperate and boreal forests is indisputable. In almost all temperate and boreal countries still possessing substantial old-growth forests, the timber trade is now the primary cause of natural forest loss.

THE TIMBER TRADE AND AREAS OF BIOLOGICAL RICHNESS

This leads to what is perhaps the most important analysis in our study, the comparison of commercial logging operations with areas of high biodiversity.

Norman Myers has identified a series of 'biological hotspots' around the world, which he argues contain particularly high levels of biodiversity, and therefore should be given priority for conservation. Some critics have questioned his choices, suggesting that other less carefully studied areas may prove to have equal levels of biodiversity. However, the general principle that biodiversity is unevenly spread around the world, and around forest areas, is not open to doubt. Human intervention has increased this 'patchiness' of biodiversity distribution.

We have applied this principle to an examination of the impacts of the timber trade. In general terms, the Earth currently contains fairly large areas of recently cleared forest, young regenerating forest and middle-age forest; ie habitats for colonizer and unspecialized species. Far less common, particularly in the North but increasingly also in the South, are *old-growth forests*. These generally have a specialized flora and fauna that can *only* survive in forest ecosystems that have developed relatively undisturbed over hundreds of years.

In many of these areas, the timber trade remains, or has become, a primary agent of change. Using the data assembled in Chapter 4 (and particularly in Table 4.2) we have identified a range of key forest ecosystems where the timber trade is either the sole or a highly significant cause of natural forest loss.

There is no accident in the overlap between biologically-rich forests and forests with large scale timber trade operations. Areas of high forest biodiversity tend to contain the oldest, and thus in many cases the most commercially valuable, trees. Furthermore, natural forests are often virtually unclaimed, in the stewardship of politically weak indigenous groups, or under the nominal control of the state. Forests with high biodiversity are, by their very nature, likely to draw the attention of the global timber trade.

ILLEGAL LOGGING OPERATIONS

Statistics and assessments from the industry have tended to draw on official studies of the *legal* timber trade, such as those carried out by the UN Food and Agriculture Organization (FAO). These rely in turn on government trade statistics. In fact, in some countries undergoing severe deforestation, the timber recorded by the Ministry of Forests is only a small proportion of the actual fellings and/or exports. Much illegal timber enters the international trade, with or without the knowledge of the importers. Often, illegality is tacitly accepted by the buyer. In the UK, for example, the Customs and Excise Office does not even have the right to block imports of timber even if these are known to have been cut and exported illegally.

Countries where illegal logging is having an important, and largely unquantified, impact on natural forests include (not an exhaustive list): Kenya; Zaire; Thailand; The Philippines; Cambodia; Laos; Vietnam; Indonesia; Brazil; Bolivia; Ecuador; and the Russian Federation.

CHANGING GLOBAL FOREST CONDITIONS

Time has also increased the relative impact of the timber trade, if only because deforestation from other causes has already progressed so far. Primary forest has now been reduced to fragments in many countries. As the amount of high quality, natural forest declines, and is increasingly confined to areas which are inhospitable to human settlement, the proportion of this that is damaged by the timber trade is likely to continue to grow.

Therefore, the timber trade's claims of innocence hold little water. For most of the world's biologically richest forest ecosystems, the actions of national and international timber trading are now critical.

THE WAY FORWARD

We started this book by suggesting that the timber trade is currently at a crossroads. There are certainly some optimistic signs. A substantial, and growing, section of the trade is prepared to take environmental issues seriously, and is making real efforts to change its practices. Developments such as the establishment of the Forest Stewardship Council provide a framework for changes in forest management that will have important benefits to wildlife. In turn, the environmental movement is becoming increasingly willing to work in partnership with responsible companies and industry organizations.

On the other hand, some sections of the trade are responding to the perceived 'threat' of environmentalism by resisting change and fighting back; pressuring governments and aid agencies, funding front groups to discredit the environmental lobby, cutting fast to beat planned controls, moving into areas where environmental controls are lax and delaying reforms. These timber traders can expect a rough ride in the future. As wood products from independently certified, well managed sources become increasingly common, there will be many more opportunities for organizations and the public to campaign against timber that has been extracted at a high environmental cost.

We repeat, this report is not an anti-timber document. The authors all support the use of wood from well-managed, environmentally and socially sustainable forests. Timber has many advantages over alternatives such as plastic and aluminium. We recognize that the timber trade is not the only threat to forests.

In fact, the needs of the timber trade and the environmental movement are not as far apart as industry scare-mongers like to pretend. Clearcutting an area and moving on might benefit a handful of people at the top of a timber company who cream off the majority of the profits, but it certainly doesn't benefit the workers on the ground any more than it does wildlife and the environment. Recent abandonment of worked-out concessions in countries as far apart as Côte d'Ivoire, the USA and Indonesia all bear eloquent witness to the human costs of bad forestry. Getting forest management right – for people and the environment – is in the interests of everyone. We call on the timber trade to respond positively to the challenge of forest sustainability, and to work with the environmental movement in realizing the vision of a world full of high quality forests.

Appendix
A New Definition of Forest Quality

PREFACE

Over the past four years, WWF International has been developing a definition of forest quality which seeks to reflect the full range of ecological, social and commercial interests concerned with forest management. The proposals set out below suggest a way of highlighting and assessing differences between forests, by defining a new set of *criteria of forest quality.* These criteria could be used to provide a more sophisticated assessment of the state of forests in different regions, and also in drawing up strategies for changing management and conservation practices to improve forest quality.

It should be noted that the word 'quality' itself has many different meanings, and at least three of these could be relevant here:

❑   as a standard;
❑   as a characteristic; and
❑   as a measure of excellence.

All of these can be reflected in the various draft criteria that follow.

CRITERIA FOR A NEW DEFINITION OF FOREST QUALITY

It is proposed that four broad criteria are needed for a more holistic definition of forest quality:

❑   authenticity;
❑   forest health;
❑   environmental benefits; and
❑   social and economic values.

Authenticity

A measurement of how closely a forest corresponds to the natural forest of the area and, in ecological terms, a way of defining optimal conditions for the preservation of biodiversity. Five main elements are important:

Natural composition of trees, and other flora and fauna

Natural composition does not simply refer to species, but also to genetic and ecosystem composition. Species often differ gradually over time, even if the area of woodland remains constant, as tree canopies are broken by wind, snow, disease and other factors, and as climate undergoes fluctuations. Neither does natural composition necessarily mean high variety, either in tree species or associated flora and fauna. Naturally occurring woodlands may have fewer species associated with them than disturbed or planted woodlands. Old forests are likely to contain specific groups of plants and animals that are unable to survive in alternative habitats.

Natural spatial variation of trees with respect to age, size, variety, spacing, and presence of dead and decaying timber

Natural spatial variation does not invariably mean a wide variety of ages and sizes of trees. In places where some catastrophic event has occurred, such as windthrow, fire or serious disease attack, the natural spatial variation at a stand level may be an even-aged monoculture. Temperate forests have a tendency towards small-scale uniformity, which leads in turn to a more varied landscape. Dead timber is a key component in almost all natural forest ecosystems, and one which has received far less attention in Europe than in North America.

Continuity of forest, ie the length of time that standing forest has existed on, or adjacent to, the site

In this definition, a natural forest may well have 'breaks' in which trees do not exist, for example following windthrow or fire. Although some forest ecosystems are frequently disturbed, for example the effects of fire on some boreal forests, disturbance seldom results in total loss of trees. In many forest systems, major disturbance is quite rare.

Integration of the forest into the broader landscape

Under natural conditions many temperate forests will not contain continuous forest cover. Forest edges, and the overall landscape mosaic, are thus often important aspects of forest quality.

Allowing natural catastrophes, such as fire, windblow, disease and pest infestation, to alter the canopy and composition of the forest
Although an important part of a natural system, accommodating natural disasters becomes difficult where land area is limited and fragmented. If reserve areas are large enough, disturbance such as windblow and fire are positive benefits, rather than 'problems', from an ecological perspective.

Management practices which mimic natural ecological processes
Many of these are still in their infancy and the subject of controversy. Further thought and investigation is needed on workable management practices that allow space for natural species, and natural processes. Nonetheless, important advances in our understanding of suitable management methods have been made in the last few years.

Forest health
An assessment of forest health with respect to disease and/or pollution damage. This includes three main elements:

Tree health
It is natural for there to be a proportion of sick and dying trees in a forest. However, there is also *unnatural* tree decline due to human activity. This includes impacts from air pollution, particularly ozone, occult acidic mists, soil acidification through air pollution, and the direct toxic effects of sulphur and nitrogen oxides. Tree health also suffers from impacts of management practices, use of biocides, choice of tree species, control of non-native tree diseases transmitted through timber imports, climate change, etc.

Health of forest flora and fauna
From an overall ecosystem perspective, the health of other forest wildlife may be as important as the health of trees themselves. Impacts can include air pollution damage to epiphytic lichens and mosses; losses to ground flora through management changes; decline in birds, insects and fungi requiring the presence of dead timber; losses of fire-dependent species; decline of species connected with older forest stands, etc.

Robustness in the face of changing conditions
Although the biological implications of climate change remain poorly understood, evidence to date suggests that the forests most at risk will be those already damaged by disturbance, fragmentation, pollution-related decline and the presence of exotic species on the edge of their ecological tolerance. Creating and protecting forests robust enough to withstand short- or long-term climate change is an important aspect of maintaining a healthy forest estate.

Environmental benefits

This is one of the best recognized of forest quality criteria, although some particular elements may be less well understood. It includes benefits that extend beyond the boundaries of the forest, such as:

Biodiversity and genetic resource conservation

Many aspects of biodiversity conservation are covered by the criteria for authenticity discussed above. However, it should be noted that in areas where native woodlands have already been severely depleted or eliminated, and conservation is based around a managed landscape, it may be necessary to preserve unnatural forest areas if specific rare species are to be protected.

Soil and watershed protection

This criterion embraces both potential positive and negative effects. Bad forestry can create additional soil erosion and siltation, upsetting hydrological cycles. Forestry in a polluted atmosphere can concentrate various air pollutants leading to problems of freshwater acidification. On the other hand, well managed forests can have a positive impact on both soil and water quality through reducing erosion, maintaining water purity, etc.

Impact on other natural or semi-natural habitats, including impacts through management and afforestation

In regions where the net area of forestry is being extended, often to counter past deforestation, management should involve assessment of what is lost when forests are established.

Local climatic benefits

Comparatively little is known about the impacts of forests on local climate conditions in temperate countries, although transpiration by trees certainly has effects on the hydrological cycle.

Carbon sequestration and climate stabilization

The medium term effectiveness of growing trees to collect carbon depends very much on what the timber or pulp is used for; many uses, such as paper and card, mean that the manufactured material may quickly break down, in which case the carbon dioxide is released into the atmosphere once again. This is particularly true if waste products are incinerated. In natural forests, the bulk of carbon is stored in the humus layer below the soil surface, which tends to be released if these are felled and replanted.

Other social and economic values

A measurement of economic and social criteria relating to forest areas, including a broad array of values to humans. These range from commercial considerations through to spiritual and religious values.

Wood products including timber, pulp and fuelwood

Timber values change over time, both in real terms and because, as uses change, different species and timber qualities are required. In this respect, quality of timber from a manufacturing perspective is also a significant aspect of overall forest quality.

Non-timber products including fruit, nuts, forage, game animals, medicines, etc

On the whole, non-timber products play a minor economic role in temperate forestry, although collection of nuts, fruits and fungi may be important from a local economic or cultural perspective.

Employment in the forest and in surrounding areas

Employment includes both direct work in the forest and indirect labour through support workers, service industries, etc. At best, forestry can bring new skills into an area and help to maintain traditional skills. In addition, the forest sometimes supplies employment through use of the land area under commons agreements and grazing rights. Well-managed forestry operations also provide growing employment in the leisure and tourism industries.

Recreation, such as walking, picnics, camping and sport hunting

Recreational needs are an increasingly important facet of forest services. Accommodating different activities in forests requires regional planning and some compromise. Different recreational activities have different 'carrying capacities' for a forest; for example a wilderness area has a very low human carrying capacity, or it no longer retains the status of wilderness, whereas a park can have a very high carrying capacity.

Homeland for people, particularly indigenous people

This is not usually regarded as a major factor in Europe. However, many people born in woodland areas, or those with a strong interest in the land, rely on forests for a proportion of their livelihoods. Many others choose to live in or near forests because of the benefits that the forests provide.

Historical importance

Many forests contain important historical sites, such as prehistoric remains, or more recent architectural and industrial heritage. Furthermore, in some forests the system of management has an historical significance of its own.

Aesthetic values
A forest's attractiveness is of key importance from its recreational point of view, and for the people living nearby. Factors which have an impact on aesthetic perceptions of the forest include age and diversity of trees, other wildlife, quality of light, quietness, forest mosaic, presence of water and many more.

Local distinctiveness and cultural values
People living in the forest on a regular basis often have different ideas about aesthetic values of forests from those held by visitors or planners. Incorporating local cultural values into forest management is important if the needs of these people are to be addressed.

Mystic and religious significance
Some cultures and religions, such as the Buddhist and Hindu faiths, have particularly strong links with forests, and all major religious philosophies acknowledge the importance of forests and other natural ecosystems.

Educational value, including the role of forests in scientific research
This includes both formal education for school and further education students, and also education in a broader sense of providing information to the general public.

Placing forest quality in a suitable framework
Many of the criteria of forest quality suggested above also need to be fitted into a framework of time and space and identified with certain interest groups. Although the following are not criteria they are important *qualifying conditions* to the criteria.

❑ **Spatial considerations:** whether quality criteria are aimed at satisfying the needs of local communities, national interests or international requirements.
❑ **Temporal issues:** whether management decisions are being taken with a regard to quality for current or future generations.
❑ **Human and non-human matters:** whether quality is seen purely from a human perspective, and the extent to which it also encompasses the needs of other living organisms.

RECONCILING DIFFERENT CRITERIA OF FOREST QUALITY

Some criteria are difficult to reconcile, and there may even be conflicts within a category. It would be virtually impossible for one medium-sized forest to be rated high for *all* criteria, and attempts to achieve this would probably result in forests that did not fully satisfy any criteria. However, **it is possible for a forest region, or a national forest policy, to reflect all aspects of the criteria of forest policy.** The weighting given to various elements will be a matter of continuing debate.

Main source: Dudley, N, J-P Jeanrenaud and S Stolton [ed] (1993) *A New Definition of Forest Quality* WWF UK, Godalming, Surrey

Notes and References

CHAPTER 2

1   Salati, E and P B Vose (1983) 'Depletion of Tropical Rainforests' *Ambio* 12 (2)
2   FAO (1986) *Informe preliminar Examen y analisi de las politicas y estrategias para el desarrollo rural en Panama*, Rome
3   Hamilton, L S and A J Pearce (1988) 'Soil and Water Impacts of Deforestation', in *Deforestation Social Dynamics in Watersheds and Mountain Ecosystems*, ed J Ives and D C Pitt, Routledge, London
4   Lean, G, P Ghazi and M Lean (undated) WWF UK supplement to *The Observer*
5   Centre for Science and Environment (1987) *The Wrath of Nature: The Impact of Environmental Destruction on Floods and Droughts*, CSE, New Delhi
6   Grainger, A (1990) *The Threatening Desert: Controlling Desertification* Earthscan, London
7   Wilson, E O (1986) *Biodiversity* National Academy of Sciences Press, Washington DC
8   Prescott-Allen, R (1992) *Caring for the Earth: A Strategy for Sustainable Living* Earthscan, London
9   Friends of the Earth (1990) *Memorandum Submitted by Friends of the Earth to the House of Commons Select Committee Enquiry into the Climatological and Environmental Effects of the Destruction of the Tropical Rainforests* HMSO, London
10 Myers, Norman (1988) 'Threatened Biotas: 'Hotspots' in Tropical Forests' *The Environmentalist* **8** (3)
11 Lattin, J D (1990) 'Arthropod Diversity in Northwest Old Growth Forests' *Wings: Essays on Insect Conservation* Xerxes Society, Summer 1990
12 Figures for Cameroon study from Ruitenbeck, J (1990) *Economic Analysis of Tropical Forest Conservation Initiatives: Examples from West Africa* WWF UK, Godalming, and for Finnish berry picking (multiple references) from Hyttonen, Marjatta (1992) *Multiple-use Forestry Research in Finland 1970–1990: An Annotated Bibliography*, Finnish Forest Research Institute, Research Paper 430, Helsinki
13 Groombridge, B [ed] (1992) *Global Biodiversity: Status of the Earth's Living Resources* World Conservation Monitoring Centre and Chapman and Hall, Cambridge and London
14 Ponting, C (1990) *A Green History of the World* Penguin, Harmondsworth, Middlesex
15 World Resources Institute (1992) *World Resources 1992–93* Oxford University Press, Oxford; also Prescott-Allen, R (1992) *Caring for the Earth* Earthscan, London
16 Jeanrenaud, J-P (1992) 'The Global Forest Crisis' Information sheet from WWF-UK, Godalming
17 Forest Resources Assessment 1990 Project (1992) *Third Interim Report on the State of Tropical Forests* Food and Agricultural Organization, June 1992
18 Jeanrenaud, J-P (1992) *op cit* Ref 16
19 World Resources Institute (1992) *World Resources 1992–93* Oxford University Press, Oxford
20 UNECE/FAO [United Nations Economic Commission for Europe and the Food and Agricultural Organisation of the United Nations] (1985) *The Forest Resources of the ECE Region* (Europe, the USSR, North America)

21 Ibero, C (1994) *The Status of Old Growth and Semi-Natural Forest in Western Europe* WWF
    International, Gland
22 Statistics in this paragraph calculated from various sources, given in Dudley, N (1992)
    Forests in Trouble: A Review of the Status of the World's Temperate Forests WWF International,
    Gland, and from Noss, R F Report for the National Biological Association of the USA
23 World Resources Institute (1992) *World Resources 1992–93* Oxford University Press,
    Oxford; Dudley, N (1992) *op cit* Ref 22
24 Some attempts to reach agreement on this in the UK context are contained in Dudley, N,
    J-P Jeanrenaud and S Stolton (1993) *Towards a Definition of Forest Quality* WWF UK,
    Godalming, report of a colloquium looking at this issue
25 UNECE/CEC (1993) *Forest Conditions in Europe: Results of the 1992 Survey* Brussels and
    Geneva
26 Roser, D J and A J Gilmour (1995) *Acid Deposition and Related Air Pollution: Its current extent
    and implications for biological conservation in Eastern Asia and the West Pacific* WWF
    International, Gland
27 Houghton, J T, G J Jenkins and J J Ephraums (1990) *Climate Change: The IPCC Scientific
    Assessment* Cambridge University Press, Cambridge; and Houghton, J T, B A Callander
    and S K Varney (1992) *Climate Change 1992: The Supplement Report to the IPCC Scientific
    Assessment* Cambridge University Press, Cambridge
28 Franklin, J F, F J Swanson, M E Harmon, D A Perry, T A Spies, V H Dale, A McKee, W
    K Ferrell, J E Means, S V Gregory, J D Lattin, T D Scholwalter and D Larsen (1992)
    'Effects of Global Warming on Forests in Northwestern America' *The Northwest
    Environment Journal*, **7**
29 Solomon, A N (1993) 'Management and Planning of Terrestrial Parks and Reserves
    During Chronic Climate Change' in J Pernetta [ed], *Impacts of Climate Change on Ecosystems
    and Species: Implications for Protected Areas* IUCN, Gland
30 Markham, A, N Dudley and S Stolton (1993) *Some Like It Hot: Climate Change, Biodiversity
    and the Survival of Species* WWF International, Gland
31 United Nations (1990) *Forest Fire Statistics 1985–1988* United Nations Economic
    Commission for Europe and Food and Agricultural Organisation, UN, New York
32 United Nations (1990) *op cit* Ref 31, and Arden-Clarke, C (1991) *Mediterranean Forest Fires*
    briefing paper from the World Wide Fund for Nature, Gland
33 Cotchell Pacific (1987) *The Cotchell Report on China*, Hong Kong; and Salisbury H E (1989)
    The Great Dragon Fire Little, Brown and Co, Boston and London
34 Cox, R and N M Collins (1991) 'Indonesia' in Collins, N M, J A Sayer and T C
    Whitmore *The Conservation Atlas of Tropical Forests: Asia and the Pacific* Macmillan Press Ltd,
    London and Basingstoke; and Goldammer, J G and B Seibert (1992) 'The Impact of
    Droughts and Forest Fires on Tropical Lowland Rain Forest of East Kalimantan', in
    Goldammer, J G *Fire in the Tropical Biota – Ecosystem Processes and Global Challenges* Springer-
    Verlag, Berlin
35 Bourke, G (1987) 'Côte d'Ivoire's dwindling forests' *West Africa* **24**, 24 Aug 1987
36 Centeno, J C (1993), paper on population for WWF International
37 Mombiot, G (1991) *Amazon Watershed* Michael Joseph, London; ‹oyal Commission (1979)
    Distribution of Income and Wealth Report no 7, Cmnd 7597, HMSC, London
38 Brown, L R and E C Wolf (1984) *Soil Erosion: Quiet Crisis in the World Economy* Worldwatch
    Paper no 60, Worldwatch Institute, Washington DC
39 Secrett, C (1989) 'The Environmental Impact of Transmigration' *The Ecologist* **16** (2/3)
40 Repetto, R (1988) *The Forest for the Trees: Government Policies and the Misuse of Forest Resources*
    World Resources Institute, Washington DC
41 WWF, IUCN and UNEP (1982?) *The World Conservation Strategy* WWF International,
    Gland
42 Utting, P (1991) *The Social Origins and Impact of Deforestation in Central America* UN Research
    Institute for Social Development, Discussion Paper 24, Geneva

43 Leach, G and R Mearns (1988) *Beyond the Woodfuel Crisis: People, Land and Trees in Africa* Earthscan, London

44 Barbier, E, J Burgess, J Bishop, B Aylward and C Brown (1993) *The Economic Linkages Between the International Trade in Tropical Timber and the Sustainable Management of Tropical Forests* report to ITTO from the London Environmental Economics Centre

45 For references see for example Hurst, P (1990) *Rainforest Politics* Zed Press, London; Hecht, S and A Cockburn (1989) *The Fate of the Forest Developers, Destroyers and Defenders of the Amazon* Verso, London; Dudley, N (1989) *Transnational Companies and Tropical Rainforests* report to Friends of the Earth from Earth Resources Research, London; Kimmerling, J (1992), report for the Natural Resources Defense Council, Washington DC

CHAPTER 3

1 UNEP (1991) *Environmental Data Report* Blackwell, Oxford

2 Durning, A T (1992) *How Much is Enough?* Earthscan, London, in association with the Worldwatch Institute, Washington DC

3 Hammond, A L [ed-in-chief] (1994) *World Resources 1994–95: A Guide to the Global Environment* World Resources Institute in collaboration with Oxford University Press, Oxford and New York

4 Hammond, A L [ed-in-chief] (1994) *op cit* Ref 3

5 Hammond, A L [ed-in-chief] (1994) *op cit* Ref 3

6 Durning, A T (1992) *op cit* Ref 2

7 *Timber Trades Journal* (1992) 25 Jan 1992

8 *Timber Trades Journal* (1992) 8 Feb 1992

9 Peakman, J K [ed] (1991) *Australia's Top One Hundred: 1991* Australia Stock Exchange Ltd

10 Peakman, J K [ed] (1991) *op cit* Ref 9

11 Timber Trades Journal (1992) 8 Feb 1992

12 Parkes, C (1988) 'An Era of Structural Change Beckons Pulp and Paper Makers', *Financial Times* 7 June 1988, London; and Payne, M (1991) *World Pulp and Paper Profiles of 150 Leading Manufacturers* EIU Special Report 2132, Economic Intelligence Unit, London

13 O'Hara, H (1991) *Forests in Crisis: The Myth of Sustainable Forestry* The Women's Environmental Network, London

14 *Pulp and Paper International* (1994) 'Top 150 Listing' *Pulp and Paper International*, Sept 1994, Belgium

15 *Timber Trades Journal* (1992) 8 Feb 1992

16 personal communication from Kymmene staff in Finland, May 1994

17 *Timber Trades Journal* (1994) 22 Oct 1994, p 4

18 Statistics on central European forests from Rotbergs, U (1994) *Forest Protection and Management in Latvia*; Vancura, K (1994) *Biodiversity of the Forest Ecosystems and Sustainable Forest Management*; Holdampf, G (1994) *Sustainable Forestry in Hungary*; and Radu, S (1994) *Forest Status in Romania*; all papers given at a Workshop on Conservation of Forests in Central Europe, organized by WWF International, the Ministry of Agriculture of the Slovak Republic and the Forest Research Institute, Zvolen, in Zvolen, Slovakia, 7–9 July 1994

19 Birchfield, R J and I F Grant (1993) *Out of the Woods: The Restructuring and Sale of New Zealand's State Forests* Q. P. Publications, Wellington

20 Foreign Control Watchdog (1993) *Campaign Against Foreign Control of Aotearoa*, Christchurch, New Zealand

21 Anon (1994) *Earth Island Journal* **9** (3), quoting the report *New Zealand Forest Products and Business Opportunities*

22 Anon (1991) *Stora Facts and Figures: 1991* Stora, Falun, Sweden

23 UNCTC (1985) *Environmental Aspects of the Activities of Transnational Corporations: A Survey* United Nations, New York

24   Nectoux, F and N Dudley (1987) *A Hard Wood Story: Europe's Involvement in the Tropical Timber Trade* Friends of the Earth, London

25   Wagner, B (1988) 'An Emerging Corporate Nobility? Industrial Concentration of Economic Power on Public Timber Tenures' *Forestry Planning Canada* **4** (2), 14–18, Victoria, British Columbia

26   Nawikta Resource Consultants (1991) data on the corporate concentration of harvesting rights and ownership in the BC forest industry, *Forest Planning Canada* **7** (4), July–Aug 1991, Victoria, British Columbia

27   Rice, T and S Counsell (1993) *Forests Foregone: The European Community's Trade in Tropical Timbers and the Destruction of the Rainforests* Friends of the Earth, London

28   UNCTC (1991) *Transnational Corporations and Issues Relating to the Environment Including the Contribution of the Commission and the United Nations Centre on Transnational Corporations to the World of the Preparatory Committee for the United Nations Conference on Environment and Development* E/C10/1991/3 25 March 1991, Economic and Social Council, United Nations, New York

29   Dudley, N (1992) *Transnational Companies and Forests* report for WWF UK prepared by Earth Resources Research, contains information on interviews with TNCs regarding standardization of procedures

30   Reuters wire service, 16 Oct 1992, quoted in *World Rainforest Report* **9** (2), World Rainforest Movement, Rainforest Action Network, San Francisco

31   Jackson, J K (1986) *A Study in Timber Transfer Pricing in PNG* UN Centre on Transnational Corporations, New York, 1986

32   Commission of Inquiry, Interim Report no 4, vol 1, p 85 'New Ireland'

33   Commission of Inquiry Interim Report no 4, 'Timber Exploitation in New Ireland Province' vol 3, appendix 6

34   Marshall, G (1990) *The Barnett Report: A Summary of the Report of the Commission of Inquiry into Aspects of the Timber Industry in Papua New Guinea* Asia Pacific Action Group, Hobart, Tasmania

35   United Nations (1991) *Preparations for the United Nations Conference on Environment and Development on the Basis of General Assembly Resolution 44/228 and taking into Account other relevant General Assembly Resolutions: Conservation and Development of Forests Report by the Secretary-General of the Conference* A/CONF151/PC/64, 10 July 1991

36   Canadian Forest Service (1994) *The State of Canada's Forests 1993* Forestry Canada, Ottawa

37   Anon (1993) *The State of Canada's Forests 1992* Forestry Canada, Ottawa; *Compendium of Canadian Forestry Statistics 1992* Canadian Council of Forest Ministers, Ottawa 1993; and also from the *FAO Forest Products Yearbook 1991*

38   Statistics on the Finnish forest industry presented by A Ruenala and C von Ungern-Sternberg to a field trip of UK timber retailers and WWF UK, Helsinki, May 1994

39   Resource Assessment Commission (1992) *Forest and Timber Inquiry Final Report Overview*, Australian Government Publishing Service, Canberra; and Australian Bureau of Agricultural and Resource Economics (1993) *Quarterly Forest Products Statistics*, Canberra

40   Anon (1992) *Country Report no 3: South Korea*, Economist Intelligence Unit, London; and Anon (1991) *What You Can Sell to Taiwan ROC*, Chinese External Trade Development Council

41   Bayliss, M (1992) *Pulp, Paper and Wood Resources in South East Asia* EIU Special Report 2554, Economist Intelligence Unit, London

42   Dudley, N and S Stolton (1994) *The East Asian Timber Trade and its Environmental Implications* WWF UK, Godalming

43   Hong Kong and Shanghai Banking Corporation (1990) *Business Profiles Service* Republic of Korea, Hong Kong

44   Food and Agriculture Organization of the United Nations (1991) *FAO Yearbook of Forest Products 1989* Rome; and Bayliss, M (1992) *op cit* Ref 41

45   Mather, A S (1990) *Global Forest Resources* Belhaven Press, London

46 McDermott, M and S Young (1989) *South Korea's Industry: New Directions in World Markets* EIU Special Report no 2005, July 1989, London
47 Bayliss, M (1992) *op cit* Ref 41
48 Anon, China External Trade Development Council (1991) *op cit* Ref 40
49 Bayliss, M (1992) *op cit* Ref 41
50 *Asian Timber* (1992) March
51 Bayliss, M (1992) *op cit* Ref 41
52 Hurst, P (1990) *Rainforest Politics: Ecological Destruction in South-East Asia* Zed Books, London
53 Dudley, N (1992) *Forests in Trouble* WWF International, Gland
54 Anon, China External Trade Development Council (1991) *op cit* Ref 40
55 Information from a Workshop on Conservation of Forests in Central Europe, organized by WWF International, the Ministry of Agriculture of the Slovak Republic and the Forest Research Institute, Zvolen, 7–9 July 1994, in Zvolen, Slovakia
56 FAO (1993) *FAO Yearbook of 1991* Forest Products, Rome
57 Arden-Clarke, C (1990) *Conservation and the Sustainable Management of Tropical Forests: The Role of ITTO and GATT* WWF International, Gland
58 Wood Supply Research Group (1991)
59 Rasmusson, U (1991), personal communication from WWF Sweden
60 Forestry Canada (1991) *The State of Forestry in Canada: 1990 Report to Parliament* Ministry of Supply and Services Canada, Ottawa
61 Tompkins, S (1989) *Forestry in Crisis: The Battle for the Hills* Christopher Helm, London
62 Anderson, H M and J T Olsen (1991) *Federal Forests and the Economic Base of the Pacific Northwest: A Study in Regional Transitions* The Wilderness Society, Washington DC
63 Egan, T (1993) 'Land Deal Leaves Montana Logged and Hurt' *New York Times*, 19 Oct 1993
64 Sedjo, R A and K S Lyon (1990) *The Long-Term Adequacy of World Timber Supply* Resources for the Future, Washington DC
65 Food and Agricultural Organization of the United Nations and the United Nations Economic Commission for Europe (1986) *European Timber Trends and Prospects to the Year 2000 and Beyond* ECE/TIM/30, United Nations, New York
66 Kroesa, R (1990) *The Greenpeace Guide to Paper* Greenpeace International, Vancouver
67 Nectoux, F (1992), in a paper to a WWF European Forests Workshop in Gland
68 Nectoux, F (1992), *op cit* Ref 67
69 Hadfield, P (1992) 'Pressure-cooked Wood is a Square Deal' *New Scientist* 4 July 1992, London

CHAPTER 4

1 personal communication to N Dudley, Oct 1991
2 Friends of the Earth (1990) *Memorandum Submitted by Friends of the Earth to the House of Commons Select Committee Enquiry into the Climatological and Environmental Effects of the Destruction of the Tropical Rainforests* HMSO, London
3 Wilson, E O (1989) 'Threats to Biodiversity' *Scientific American*, Sept 1989
4 Whitmore, T C and J A Sayer (1992) *Tropical Deforestation and Species Extinction* IUCN, Gland
5 Schowalter, T D (1989) 'Canopy Arthropod Community Structure and Herbivory in Old-growth and Regenerating Forests in Western Oregon' *Canadian Journal of Forest Research* **19**, pp 318–22
6 Moore, P at a meeting with the WWF in October 1992
7 Brown, E R [technical ed] (1985) *Management of Wildlife and Fish Habitats in Forests in Western Oregon and Washington* USDA Forest Service, Pacific Northwest Region, Portland, Oregon

8   Wilderness Society (1988) *End of the Ancient Forests: Special Report on National Forest Plans in the Pacific Northwest* The Wilderness Society, Washington DC

9   Norse, E A (1990) *Ancient Forests of the Pacific Northwest* Island Press for the Wilderness Society, Washington DC

10   Denison, W C (1990) 'Fungi of the Ancient Forests: Hundreds of Mushrooms, Legions of Molds' in Norse, E A (1990) *op cit* Ref 9

11   Trappe, J M (1990) 'The "Most Noble" Polypore Threatened' in Norse, E A (1990) *op cit* Ref 9

12   Schowalter, T D (1989) *op cit* Ref 6

13   Wilcove, D S (1988) *National Forests: Policies for the Future: vol 2, Protecting Biological Diversity* The Wilderness Society, Washington DC

14   Durbin, K (1991) 'Fishery Experts say Forest Plans put Fish at Risk' *The Oregonian*, 29 Oct 1991, Portland

15   Nehlsen, W, Williams, J E and Lichatowich, J A (1991) 'Pacific Salmon at the Crossroads: stocks at risk from California, Oregon, Idaho and Washington' in *Fisheries* **16** (2), pp 4–21

16   Della Salla, D personal communication to N Dudley, 1995

17   Durbin, K and P Koberstein (1990) 'Survival Hinges on Old growth Habitat' *Special Report: Forests in Distress, The Oregonian*, 15 Oct 1990, Portland, Oregon

18   Gamlin, L (1988) 'Sweden's Factory Forests' *New Scientist* 28 Jan 1988, London

19   Elmberg, J (1991) personal communication

20   Franklin, J F and R T T Foreman (1987) 'Creating Landscape Patterns by Forest Cutting: Ecological Consequences and Principles' *Landscape Ecology* **1** (1), pp 5-18

21   Simberloff, D (1992) 'Do Species-are Curves Predict Extinction in Fragmented Forest?' in *Tropical Deforestation and Species Extinction* ed by T C Whitmore and J A Sayer, IUCN, Gland

22   DellaSala, D A: Olsen, D M: Barth, S E: Crane, S L: Primm, S A *Forest Health: moving beyond the rhetoric to restore healthy landscapes in the island northwest* (in press)

23   Amnesty International (1988) *Brazil: Cases of Killings and Ill-Treatment of Indigenous People* London

24   Survival International, personal communication, March 1994

25   Survival International, personal communication, March 1994

26   Utting, P (1991) *The Social Origins and Impact of Deforestation in Central America* UN Research Institute for Social Development, Discussion Paper 24, Geneva

27   O'Hara, H (1991) *Forests in Crisis: The Myth of Sustainable Forestry* The Women's Environmental Network, London

28   Salati, E and P B Vose (1983) 'Depletion of Tropical Rainforests' *Ambio* **12** (2)

29   FAO (1986) *Informe preliminar Examen y analisi de las politicas y estrategias para el desarrollo rural en Panama* Rome

30   Anderson, H M and C Gehrke (1988) *National Forests: Policies for the Future: volume 1, Water Quality and Timber Management* The Wilderness Society, Washington DC

31   Adams, P W, A L Flint and R L Fredriksen (1991) 'Long-Term Patterns in Soil Moisture and Revegetation after a Clearcut of a Douglas-Fir Forest in Oregon' *Forest Ecology and Management* **41**, pp 249–63

32   World Paper (1994) 'Companies Face Charges' *World Paper*, vol 219 (5), June 1994

33   Lean, G, P Ghazi and M Lean (undated, 1992), supplement to *The Observer*

34   Brown, G (1980) *Forestry and Water Quality* Oregon State University Book Stores Inc, Corvalis, Oregon

35   Grant, G (1990) 'Hydrologic, Geomorphic and Aquatic Habitat Implications of Old and New Forestry' *Forests – Wild and Managed: Differences and Consequences* [ed] A F Pearson and D A Challenger, 19–20 Jan 1990, University of British Columbia

36   Lean, G, P Ghazi and M Lean (undated, 1992), supplement to *The Observer* and Agarwal, A *et al* (1987) *The Wrath of Nature* Centre for Science and the Environment, Delhi

37   Hall, D O and F Rosillo-Calle (1989) CO_2 *Cycling by Biomass: Global Productivity and Problems*

of Deforestation paper presented at the conference 'Amazonia, Facts, Problems and Solutions' 31 July – 2 Aug, University of Sao Paulo, Brazil
38   Muller, F (1992), presentation to the World Bank, Washington DC
39   Hamilton, L S and A J Pearce (1988) 'Soil and Water Impacts of Deforestation' *Deforestation: Social Dynamics in Watersheds and Mountain Ecosystems* ed J Ives and D C Pitt, Routledge 1988
40   Nicholson, D I (1979) 'The Effects of Logging and Treatment on the Mixed Dipterocarp Forests of South East Asia' FAO Report FOMISC/79/8, Food and Agricultural Organization, Rome
41   Perry, D A (1988) 'Landscape Pattern and Forest Pests' *The Northwest Environmental Journal* **4**, pp 212–28, University of Washington, Seattle
42   Phipps, A (1991) personal communication from the Alaska Center for the Environment
43   Sierra Club of Western Canada (undated) *BC Forest Log Sheet*, Vancouver
44   Global Witness (1995) *Forests, Famine and War: The key to Cambodia's future* a briefing document by Global Witness, 9 March 1995
45   Nectoux, F (1985) *Timber!* Friends of the Earth, London; Nectoux, F and N Dudley (1987) *A Hard Wood Story: Europe's Involvement in the Tropical Timber Trade* Friends of the Earth, London; Nectoux, F and Y Kuroda (1989) *Timber from the South Seas* WWF International, Gland; Rice, T and S Counsell (1993) *Forests Foregone: The European Community's Trade in Tropical Timbers and the Destruction of the Rainforests* Friends of the Earth, London; internal papers on the US timber trade from the Rainforest Action Network, San Francisco and Greenpeace USA, Washington DC
46   Dudley, N and S Stolton (1993) *The East Asia Timber Trade* WWF UK, Godalming, Surrey
47   Figures calculated from Callister, D (1992) *Illegal Tropical Timber Trade: Asia-Pacific* Traffic International, Cambridge, UK; see also Dudley, N (1991) *The Impact of Thailand's Logging Ban on Deforestation* Earth Resources Research for WWF UK, London
48   Mombiot, G (1991) *Amazon Watershed: The New Environmental Investigation* Michael Joseph, London; see also Read, M (1990) *Mahogany: Forests or Furniture* Fauna and Flora Preservation Society, Brighton; Greenpeace (1993) *Mahogany* leaflet from Greenpeace UK, London; Greenpeace (1993) *Predatory Mahogany Logging: A Threat to the Future of the Amazon* Greenpeace UK, London; Tickell, O (1993) 'A Saw Point in Brazil' *The Guardian* 15 Oct 93; Snook, L K (1994) *Mahogany: Ecology, Exploitation, Trade and Implications for CITES* report for WWF US, Oct 1994; Bonner, J (1994) 'Battle for Brazilian Mahogany' *New Scientist*, 22 Oct 1994, London
49   Read, M (1993) *Ebonies and Rosewoods: Requiem or Revival?* Fauna and Flora Preservation Society, London
50   Friends of the Earth (1992) *Plunder in Ghana's Rainforest for Illegal Profit: An Exposé of Corruption, Fraud and Other Malpractices in the International Timber Trade* FoE, London
51   Marshall, N T and M Jenkins (1994) *Hard Times for Hardwoods: Indigenous Timber and the Timber Trade in Kenya* TRAFFIC East/Southern Africa, Cambridge (UK)
52   Taiga Rescue Network (1993) *The Taiga – a Treasure – or Timber and Trash?* Taiga Rescue Network, Jokkmokk, Sweden
53   Anderson, P *et al* (1993) *Quick Cash for Old-Growth: The Looting of Russia's Forests* Greenpeace International, Amsterdam
54   Barr, B M (1988) 'Perspectives on Deforestation in the USSR' in *World Deforestation in the Twentieth Century* ed J F Richards and R P Tucker, Duke University Press, Durham, NC and London
55   FAO [Food and Agriculture Organization] (1991) *Forest Products Yearbook 1989* FAO, Rome
56   Komarov, B (1978) *The Destruction of Nature in the Soviet Union* Pluto Press, London
57   Barr, B M and K E Braden (1988) *The Disappearing Russian Forest: A Dilemma in Soviet Resource Management* Rowman and Littlefield, Totowa, NJ, and Hutchinson, London
58   Loskutov, A (1993) 'Going Deeper into Russia's Forests' *Asian Timber*, June 1993
59   Grigoryev, A (1992) 'Status Report of the Forest Situation in Russia' *Taiga News* **2**,

International Working Group on Boreal Forests, Jokkmokk, Sweden
60 Wagner, H (1993) *Investing in Russia's Forests: The World Bank's Approach* presented to the International Conference in Joensuu, Finland – 'Cooperation, Planning and Financing of Forestry Projects in Russia', Sept 1993
61 Taiga Rescue Network (1993) *op cit* Ref 52
62 Neilson, D (1994) 'Vast Timber Resources but...' *Asian Timber* Jan 1994, pp 16–18
63 Rosencratz, A (1991) '...And Cutting Down Siberia' *Washington Post*, 18 Aug 1991, Washington DC
64 Quoted in Neilson, D (1994) *op cit* Ref 62
65 Krever, V, E Dinerstein, D Olsen and L Williams [eds] (1993) *Conserving Russia's Biological Diversity: An Analytical Framework and Initial Investment Portfolio* WWF US, Washington DC
66 Pisarenko (1990) quoted in Greenpeace 1993 *op cit* Ref 48
67 Isayev, A and A Yablakov (1992) 'Not to make mistakes again' *Green World* **15** (16)
68 Tysplenkov, S (1993) *The State of Forests in the Karelia and Leningrad Region* unpublished briefing document, Greenpeace Russia, Moscow
69 Pearce, M (1993) 'Rushing into Russia?' *Asian Timber* June 1993
70 Pease, D A (1992) 'Joint Venture Produces Birch Plywood in Russia' *World Wood* Oct 1992
71 The Post-Soviet Business Intelligence Digest (1992) 'Industry Wood & Pulp' *Arguments and Facts International: The Post-Soviet Business Intelligence Digest*, Aug–Sept 1992, 3:20
72 Payne, M (1991) *World Pulp and Paper: Profiles of 150 Major Manufacturers* EIU Special Report 2132, Economist Intelligence Unit, London
73 Anderson, P *et al* (1993) *op cit* Ref 48
74 Anderson, P *et al* (1993) *op cit* Ref 48
75 Anon (1992) *Timber Trades Journal* **363**, 31 Oct 1992
76 Mollersten, B (1992) *Swedish Companies in Russian and Baltic Forests* report for WWF Sweden [English summary]
77 Loskutov, A (1993) 'Going Deeper into Russia's Forests' *Asian Timber*, June 1993
78 Bradshaw, M, J (1992) *Siberia at a time of Change* EIU Special Report 2171, March 1992, Economist Intelligence Unit, London
79 Taiga Rescue Network (1993) *op cit* Ref 52
80 Bradshaw, M, J (1992) *op cit* Ref 78
81 Barr, B M (1988) *op cit* Ref 54; and Barr, B M and K E Braden (1988) *op cit* Ref 55
82 Petrov, D (1992) 'Juke Box in the Forests of Siberia: What's the Tune' article on Green Net electronic data base, London
83 The Post-Soviet Business Intelligence Digest (1992) *op cit* Ref 69
84 Petrov, D (1992) *op cit* Ref 82
85 Gordon, D (1993) 'Russian Forestry Update' *Taiga News* **5**, Taiga Rescue Network, Jokkmokk, Sweden
86 Taiga Rescue Network (1993) *op cit* Ref 50; also Grigoryev, A (1992) 'Hot Spots in Russian Forests' *Taiga News* **2**, July 1992, International Working Group on Boreal Forests, Jokkmokk, Sweden
87 *Timber Trades Journal* (1993) News, 9 Oct 93
88 Aksim, M (1993) 'Russian Forests Under Attack from all Sides' *Timber Trades Journal*, 17 April 1993
89 Greenpeace (1992) 'Rainbow Warrior Confronts Logging Barge in Russian Forests' Greenpeace press release from Environet, London
90 Gordon, D (1993) *op cit* Ref 85
91 Gordon, D (1992) 'Joint Logging Ventures in the Russian Far East' *Taiga News* **2**, International Working Group on Boreal Forests, Jokkmokk, Sweden
92 Stevens, W K (1992) 'Experts say Logging of Vast Siberian Forest could Foster Warming' *The New York Times*, 28 Jan 1992
93 Klahn, J (1991) 'USSR Timber Operations Fail to Excite Investors' *The Anchorage Times*, 3 Nov 1991, Anchorage, Alaska

94 GreenNet 13 Oct 1994, London
95 Grigoriev, A (1995) in *The Taiga Trade* ed R Olsson, Taiga Rescue Network, Jokkmokk, Sweden
96 Sved, R (1993) 'Norwegian Government Supports Logging' *Taiga News* **5**, Taiga Rescue Network, Jokkmokk, Sweden; also see Anderson, P *et al* (1993); *op cit* Ref 51
97 Rosencratz, A (1991) *op cit* Ref 63
98 Neilson, D (1994) *op cit* Ref 62
99 *Principles of the Forest Legislation of the Russian Federation* Russian Government publication, Moscow, Sept 1993
100 Russia and the Other states of the CIS Newsletter (1993) 'Russia and Former Communist Countries: Political Background, Trade Changes, Forestry' *Russia and the Other States of the CIS Newsletter*, June–July 1993, Clifford Chance Publications, London
101 The Post-Soviet Business Intelligence Digest (1992) *op cit* Ref 71
102 Thomas, B (1993) 'Facing a Market Economy' *Paper*, Oct 1993, 218
103 Grigoriev, A (1995) *op cit* Ref 95
104 Grigoriev, A (1993) 'Russia's New Forestry Act Leaving the Door Wide Open for Ruthless Exploitation' *Taiga News* **5**, Taiga Rescue Network, Jokkmokk, Sweden
105 Natural Resources Canada (1995) *The State of Canada's Froests 1994* Ottawa
106 Forestry Canada (1991) *The State of Forestry in Canada: 1990 Report to Parliament*, Ministry of Supply and Services Canada, Ottawa
107 See for example Greenpeace (1991) 'BC logging threatens forests' *Greenpeace Canada Action* **2** (3), July/Aug/Sept 1991, Toronto; Hardstaff, P (1995) *Sustainability and Logging in Canada's Forests* Briefing Sheet from Friends of the Earth, London, June 1995
108 Bourdages, J-L (1990) *Rain Forests in Canada and Brazil* Research Branch, Library of Parliament, Canada
109 Moore, P (1992) written memorandum from the BC Forest Alliance criticizing WWF's report *Forests in Trouble* and personal communication at the Canadian High Commission, Oct 1992
110 Brown R G (1993) *Regeneration Success in British Columbia's Forests* Province of British Columbia, Ministry of Forests, prepared for the 14th Commonwealth Forestry Conference, Sept 1993
111 Sierra Club of Western Canada (undated a); *BC Forestry Fact Sheet*; Sierra Club of Western Canada (undated b); *A comparison of forestry industry statistics for British Columbia and the United States* information sheet; and Travers, R (1991), *Comparative data charts* Forest Planning Canada 7 (3), Vancouver
112 Young (1991) 'More Clearcuts, Fewer Jobs' *Monday Magazine* 11–17 July 1991
113 Connely, J (1991) 'The Big Cut' *Sierra* **76** (3), pp 42–53, Sierra Club, San Francisco
114 Moore, K (1991) *Coastal Watersheds: An Inventory of Watersheds in the Coastal Temperate Forests of British Columbia* Earthlife Canada Foundation; and Ecotrust/Conservation International, Vancouver; and Roemer, H L, J Pojar and K R Joy (1988) 'Protected old growth forests in coastal British Columbia' *Natural Areas Journal* **8** (3), pp 146–58
115 Moore, K (1991) *Coastal Watersheds: An Inventory of Watersheds in the Coastal Temperate Forests of British Columbia* Earthlife Canada Foundation and Ecotrust/Conservation International, Vancouver
116 Travers, R (1991), 'Comparative data charts' *Forest Planning Canada* **7** (3), Vancouver
117 Canadian Pulp and Paper Information Centre – Europe (1995) *Clearcutting: A Canadian Perspective* Brussels
118 quoted in Western Canada Wilderness Committee (1990) 'Crisis in the Woods' *Western Canada Wilderness Committee Educational Report* **9** (8), Nov 1990, Vancouver
119 Annett, W (1991) 'Macblo: leaving home' *Canadian Business*, July 1991
120 Sierra Club of Western Canada (undated) *A comparison of forestry industry statistics for British Columbia and the United States* information sheet
121 *Financial Post* (1991) 'Phoney charges hit forest farms' editorial on 9 July 1991, Ontario

122 Forest Alliance of British Columbia (various dates, 1990s) 'What Greenpeace isn't telling Europeans' brochure, Vancouver; 'Tropical and Temperate Rainforests' brochure, 'People and Forests: A Fresh Approach' and others

123 Editorial (1993) 'Biodiversity Guidelines: Coastal Stand-Level Biodiversity and Landscape Biodiversity' *Forestry Planning Canada* **9** (1)

124 Tripp Biological Consultants Ltd (1994) *The Use and Effectiveness of the Coastal Fisheries Forestry Guidelines in Selected Forest Districts of Coastal British Columbia*, Nanaimo, British Columbia

125 WWF Canada (1993) *Making Choices*, Toronto

126 Canadian Pulp and Paper Information Centre – Europe (1995) *Forest Land Zoning in Canada and Clayoquot Sound* information sheet, Brussels

127 International Relations, Forests and Environment (1995) *Protecting Biodiversity in British Columbia: Backgrounder* Province of British Columbia, Victoria

128 The Scientific Panel for Sustainable Forest Practices in Clayoquot Sound (1995) *First Nations' Perspectives: Relating to Forest Practice Standards in Clayoquot Sound* Victoria

129 The Scientific Panel for Sustainable Forest Practices in Clayoquot Sound (1995) *A Vision and Its Context: Global Context for Forest Practices in Clayoquot Sound* Victoria

130 The Scientific Panel for Sustainable Forest Practices in Clayoquot Sound (1995) *Report 5: Sustainable Ecosystem Management in Clayoquot Sound: Planning and Practices* April 1995, Victoria

131 ibid

132 News Release (1995) 'Government adopts Clayoquot Scientific Reports Moves to Implementation' Ministry of Forests and Ministry of Environment, Lands and Parks, British Columbia, 6 July 1995, Victoria

133 Information provided by Tim Grey, Wildlands League, Ontario, May 1995

134 Marx, M (1994) 'Mitsubishi: The money and the power behind Alpac' *Boreal Forest Campaign Educational Report* **13** (7), Western Canada Wilderness Committee, Edmonton

135 Pratt, L and I Urquhart (1994) *The Last Great Forest: Japanese Multinationals and Alberta's Northern Forests* NeWest Press, Edmonton

136 McInnis, J (1994) *Japanese Investment in Alberta's Taiga Forest* presentation to the People's Forum 2001, Tokyo, Japan, quoted in Hardcastle, *op cit*

137 MacDonald, J (1990) 'Daishowa wants to talk about logging on Lubicon claim' *The Edmonton Journal*, 11 Oct 1990

138 McInnis, J (1994) 'The Great Alberta Giveaway' in *The Taiga Trade* [ed] R Olsson, Taiga Rescue Network, Jokkmokk, Sweden

139 Hammond, H (1991) *Seeing the Forest Among the Trees: The Case for Wholistic Forest Use* Polestar, Vancouver

140 Viereck, L A and E L Little (1972) *Alaska Trees and Shrubs* US Department of Agriculture Handbook no 410, US Forest Service, Fairbanks, Alaska

141 Miller, R K and A F Gasbarro (1989) *Planning a Forest Inventory: Guidelines for Managers of Alaska Native Lands* Agricultural and Forestry Experiment Station, School of Agriculture and Land Resources Management, University of Alaska Fairbanks, Oct 1989

142 Della Sola, D; personal communication to N Dudley

143 Rennick, P (1985) 'Alaska's Forest Resources' *Alaska Geographic* **12** (2), Alaska Geographic Society, Anchorage

144 ibid

145 Rennick, P (1985) *op cit* Ref 143

146 Wilderness Society (1986) *America's Vanishing Rain Forest: A Report on Federal Timber Management in Southeast Alaska* The Wilderness Society, Washington DC

147 Wilderness Society (1991) *Tongass National Forest: America's Vanishing Rainforest* fact sheet, The Wilderness Society, Washington DC

148 Juday, G P (1992) personal communication from Fairbanks University, Alaska

149 Wilderness Society (1991) *op cit* Ref 146

150 Chen, J; Franklin, F and Spies, T A; 'Microclimatic pattern and basic biological responses

at the clearcut edges of old-growth Douglas-fir stands' *New Environmental Journal* **6**, pp 424–5

151 Phipps, A (1992) Personal communication from the Alaska Centre for the Environment, Anchorage
152 Alaska Department of Natural Resources (1991) *Susitna Basin Recreation Rivers Management Plan: Newsletter* June 1991
153 Peale, M (1991) personal communication, Nov 1991
154 Lee, M (1991) personal communication from the Board of Land Management, Fairbanks, Alaska
155 Colchester, M (1995) *Suriname: The Malaysian disease: Another country tempted by timber suicide*
156 Sizer, N and R Rice (1995) *Backs to the Wall in Suriname: Forest Policy in a Country in Crisis* Earthscan, London
157 Timberlake, L (1995) 'Timber Treatment' *The Guardian*, London
158 Sizer, N and R Rice (1995) *op cit*
159 Sizer, N (1995) 'Rain Forest Doesn't Have to be Felled' *International Herald Tribune*, 17 May 1995
160 Sizer, N (1995) personal communication
161 Colchester, M (1994) 'The New Sultans of the West: Asian loggers move in on Guyana's forests', World Rainforest Movement, Penang, Malaysia and Chadlington, Oxford
162 WCMC (1991) 'Pre-Project Study on the Conservation Status of Tropical Timbers in Trade', final report to the International Tropical Timber Organization by the World Conservation Monitoring Centre, Cambridge, UK
163 Rice, T and S Counsell (1993) *Forests Foregone: The European Community's Trade in Tropical Timbers and the Destruction of the Rainforests* Friends of the Earth, London
164 Information from WWF office in Cameroon
165 Poore, D *et al* (1989) *No Timber Without Trees* Earthscan, London
166 Jeanrenaud, J-P (1990) *An Assessment of the Conservation Significance of the Tropical Forestry Action Plan for Cameroon,* unpublished report for WWF UK
167 Sayer, J S, CS Harcourt and NM Collins (1992) [eds] *The Conservation Atlas of Tropical Forests: Africa* The International Union for Conservation of Nature and the World Conservation Monitoring Centre, Macmillan Press, Basingstoke, UK
168 Rice, T and S Counsell (1993) *op cit* Ref 163
169 Buttoud, G (1991) *Les Bois Africains a l'Épreuve des Marches Mondiaux,* Ecole Nationale du Génie Rural, des Eaux et des Forêts, Nancy, France, quoted in Rice and Counsell *op cit*; and Nectoux, F and N Dudley (1987) *A Hard Wood Story* Friends of the Earth, London
170 Gartland, S (1990) *Practical Constraints on Sustainable Logging in Cameroon* Proceedings of the Conference Sur la Conservation et l'Utilisation Rationelle de la Forêt Dense D'Afrique Centrale et L'Ouest, 5–9 Nov 1990, African Development Bank, IUCN, World Bank
171 Franks, A (1990) 'An axe over nature's nursery' *The Times*, 2 March 1990, London
172 Verhagen, H and C Enthoven (1993) *Logging and Conflicts in the Rainforests of Cameroon* Milieu Defensie and Netherlands Committee for IUCN, Amsterdam
173 ibid

CHAPTER 5

1 Perlin, J (1991) *A Forest Journey: The Role of Wood in the Development of Civilization* Harvard University Press, Cambridge, Massachusetts
2 Brundtland, G H [Chair of the World Commission on Environment and Development] (1987) *Our Common Future*, Oxford University Press, Oxford
3 Gilfillan, B D, J R Cuthbert, T J Drew, L A Johnson and B J Mayes (1990) *Report of a Forestry Mission to Scandinavia* FRDA Report 156, ERDA Canada and British Columbia, March 1990

4   International Union for the Conservation of Nature (1992) *Protected Areas of the World: A Review of National Systems* vol 2, Palaearctic IUCN, Gland

5   Bain, C (1987) *Native Pinewoods in Scotland: A Review 1957–1987* Royal Society for the Protection of Birds, Edinburgh

6   Groome, H (1992) 'Pulpwood Plantation in Northwest Spain' *The Ecologist* **22** (3), May–June 1992, Sturminster Newton, Dorset

7   Economist Intelligence Unit (1991) *Chile Country Profile 1991–1992* EIU, London

8   Brown, P (1994) *The Guardian*, 15 Nov 1994, London

9   Karjalainen, H (1994) *A Finnish Forest Strategy* WWF Finland, Helsinki

10   Southwood, T R E (1961) *Journal of Animal Ecology* **30**, 1–8

11   Mathers, M, C Barden, C Hatton, J-P Jeanrenaud and F Sullivan (1992) *WWF UK Evidence to the House of Commons Select Committee on Environment Inquiry into Forestry* WWF Scotland, Aberfeldy, Perth

12   Arden-Clarke, C (1991) *Mediterranean Forest Fires,* a briefing paper from the WWF, Gland; Groome, H (1992) *Reforestation, Afforestation and Forest Fires Case Study, Spain,* paper from EHNE, Bilbao, prepared for the WWF European Forests Workshop, Gland, April 1992; Guarrera, L (1991) *Forest Fires in Italy* WWF Italy, Rome; Komarov, B (1978) *The Destruction of Nature in the Soviet Union* Pluto Press, London; Salisbury, H E (1989) *The Great Dragon Fire* Little, Brown and Co, Boston and London; and United Nations (1990) *Forest Fire Statistics 1985–1988* UNECE and FAO, United Nations, New York

13   Komareck, E V Snr (1964) 'The Natural History of Lightning' in *Proceedings of the Tall Timber Fire Ecology Conference* **3**, 139–84, Tall Timbers Research Station, Tallahassee, Florida; and Phipps, A (1991) personal communication from the Alaska Center for the Environment, Anchorage

14   DellaSala, D A; Olson, D M; Barth, S E; Crane, S L and Primm, S A (1995) 'Forest Health: moving beyond rhetoric to restore healthy forests in the inland Northwest' *Wildlife Society Bulletin*, 23 (3)

15   Dudley, N and J-P Jeanrenaud [eds] (1994) *Finland Study Tour by members of the WWF UK 1995 Group to research options for timber certification* May 23–26 1994, WWF UK, Godalming, Surrey

16   Smith, J M B and B M Waterhouse (1988) 'Invasion of Australian Forests by Woody Plants' in *Australia's Everchanging Forests* ed N J Frawley and N Semple, Special Publication no 1, Department of Geography and Oceanography, Australian Defence Force Academy, Campbell, ACT

17   Godley, E J (1975) 'Flora and Vegetation' in *Biogeography and Ecology in New Zealand* ed G Kuschel, Dr W Junke BV, Publishers, The Hague

18   Clover, C (1992) 'EC to Prosecute Britain Over Destroyed Scottish Moorland' *Daily Telegraph*, 6 Aug 1992, London

19   IUCN [International Union for the Conservation of Nature] (1992) *op cit* Ref 4

20   Tompkins, S (1988) *Forestry in Crisis: The Battle for the Hills* Christopher Helm, London

21   Robinson, M and K Blyth (1982) 'The Effect of Forestry Drainage Operations on Upland Sediment Yields: a Case Study' *Earth Surface Processes and Landforms* **7**, pp 85–90

22   Shiva, V and J Bandyopadhyay (1984) *Ecological Audit of Eucalyptus Plantations* Dehra Dun, India

23   Blockhaus, J [ed] *IUCN Forest Conservation Programme Newsletter* no 20, Nov 1994, Gland

24   Norse, E A (1990) *Ancient Forests of the Pacific Northwest* Island Press for The Wilderness Society, Washington DC

25   Swanson, F J and J F Franklin (1991) *New Forestry Principles from Ecosystem Analysis of Pacific Northwest Forests,* paper given to AAAS meeting, Washington DC, 19 Feb 1991, submitted to *Ecological Applications*

26   Adams, J A (1978) 'Long Term Aspects of Nutrient Loss from Forest Soil Ecosystems' *New Zealand Journal of Forestry* **23** (1)

27   Perry, D A (1988) 'Landscape Pattern and Forest Pests' *The Northwest Environmental Journal*

4, pp 212–28, University of Washington, Seattle

28 Radford, T (1994) 'Discovery of Source of Dutch Elm Disease "may hold key to control"' *The Guardian*, 17 Nov 1994, London

29 Purnell, S (1993) 'Nematode in the Hole Threatens Pine Imports' *Daily Telegraph*, 3 April 1993, London

30 Dudley, N (1989) *Nitrates* Green Print, London

31 Thorpe, V and N Dudley (1987) *Pall of Poison* The Soil Association, Stowmarket, Suffolk

32 Hurst, P, A Hay and N Dudley (1992) *The Pesticides Handbook* Journeyman Press, London

33 Lindahl, K [ed] (1992) *Taiga News* no 1, International Working Group on Boreal Forests, Jokkmokk, Sweden

34 Presslie, D J (1988) 'Use of Herbicides for Site Preparation and Brashing and Weeding in the Sub-Boreal Zone – A Northwood Perspective' in *Learning from the Past: Looking to the Future: Proceedings of the Northern Silvicultural Committee's 1988 Winter Workshop* ed B A Scrivener and J A MacKinnen, Feb 1989

35 Hardell, L and M Eriksson (1988) 'The Association Between Soft Tissue Sarcomas and Exposure to Pheoxyacetic Acids' *Cancer* **62**, pp 652–6

36 Rosomon, G (1994) *The Plantation Effect* Greenpeace New Zealand, Auckland

37 Dudley, N, M Barrett and D Baldock (1986) *The Acid Rain Controversy* Earth Resources Research, London

38 Newsome, M (1985) 'Forestry and Water in the Uplands of Britain: the Background of Hydrological Research and Options for Harmonious Land Use' *Quarterly Journal of Forestry* **79** pp 113–120

39 Muniz, I (1983) 'The Effects of Acidification on Norwegian Freshwater Ecosystems' in *Ecological Effects of Acid Deposition* National Swedish Environmental Protection Board report SNV PM 136, Solna, Sweden; and Tyler, S J and S J Ormerod (1986) *Dippers (Cinclus cinclus) and Grey Wagtails (Motacilla cinerea) as Indicators of Stream Acidity in Upland Waters*, ICBP World Conference, Kingston, Canada, 14 June 1986

40 Drawn partly from Bass, S (1993) 'Social Environment' *Tree Plantation Review Study no 5*, Shell International Petroleum Company and WWF, Godalming, UK; and from Dudley, N (1994) *Forests and people in Rural Areas* a discussion paper for the Forestry Commission, WWF Scotand and Scottish Natural Heritage, Edinburgh

41 *Tropical Timbers* (1988) 'Giant Takes up Forestry', Sept 1988

42 Limprungpatanakit, A (1988) 'Shell Tests Reaction to Eucalyptus Project' *Bangkok Post*, 18 April 1988, Bangkok

43 Butler, S (1987) 'Strong Pulp Prices Boost Eucalyptus Paper Profits' *Financial Times*, 2 Sept 1987, London; and Smith, M (1987) 'Ibstock Takes Full Control of Eucalyptus' *Financial Times*, 24 Nov 1987, London

44 Bernadotte, B and U Gustafsson [eds] (1987) *Swedish Forest: Facts about Swedish Forestry and Wood Industries* Skogsstyrelsen [National Board of Forestry], Jönköping

45 Gamlin, L (1988) 'Sweden's Factory Forests' *New Scientist*, 28 Jan 1988, London

46 Bernes, C (1994) *Biological Diversity in Sweden: A Country Study*, Swedish Environmental Protection Agency, Solna

47 Tuvesson, A (1992) personal communication

48 Bernadotte, B and U Gustafsson [eds] (1987) *op cit* Ref 44

49 Tjernberg, M (1986) *The Golden Eagle and Forestry* Swedish University of Agricultural Science, Department of Wildlife Ecology, Report no 12, Uppsala

50 Nitare, J (1991) 'Nyckelbiotoper avslojar skogens historia', *Skogs-Eko* April 1991, pp 12–13

51 Bernadotte, B and U Gustafsson [eds] (1987) *op cit* Ref 44

CHAPTER 6

1 Figures calculated from Food and Agriculture Organization (1993) *FAO Yearbook of Forest Products 1991* FAO, Rome

2   Kroesa, R (1990) *The Greenpeace Guide to Paper* Greenpeace International, Vancouver, Canada
3   Nectoux, F (1992) in a paper to a WWF European Forests Workshop, Gland
4   Chemprojects Design and Engineering Prt Ltd (1992) *Indian Scenario for Raw Materials for the Pulp and Paper Industry 1991–2 to 2010* New Delhi, India
5   Uutela, E (1989) 'Strong Growth Predicted to Year 2000' *Pulp and Paper International*, Jan 1989, London
6   Canadian Pulp and Paper Industry (1991) *Statistics of World Demand and Supply* Montreal
7   Pulp and Paper International (1988) quoted in Kroesa, R (1990) *op cit* Ref 2
8   Matussek, H, W Salvesen and J Pearson (1993) *Pulp and Paper International Fact and Price Book* Miller Freeman, USA
9   Food and Agriculture Organization, annual timber statistics 1993
10   *Business Korea* (1990) 'Publications Mushroom Stimulating Paper Industry' **7** (8), Feb 1990
11   Anon (1993) 'Pulp and Paper in Romania' *Paper Technology* **34** (3), pp 12–13
12   Shevchenko, S M (1994) 'Pulp and Paper Between Peace and War' *Tappi J* 77 (1), pp 61–6
13   Ljunggren, A (1994) 'The Changing Landscape of Eastern Europe' *International Papermaker* **57** (5), pp 20–23
14   FAO (1993) *Pulp and Paper Capacities: Survey 1992–1997* FAO, Rome
15   Figures taken from VDP (1986), the West German Industry Association, quoted in Kroesa, R (1990) *op cit* Ref 2; Elkington, J and J Hailes (1988) *The Green Consumer Guide* Gollancz, London
16   Calculated from Food and Agriculture Organization (1993) *FAO Yearbook of Forest Products 1991* FAO, Rome
17   Figures calculated from data in Food and Agriculture Organization (1994) *Yearbook of Forest Products* FAO, Rome
18   Dudley, N (1992) *The Price of Preservation: New Zealand's Plantation Policy and Preservation of Primary Forests* Earth Resources Research, London
19   Rasmusson, U (1994) *Swedish/Scandinavian involvement in Indonesian forestry – The Industrial Forest Plantations and Pulp Mill Sector* WWF Sweden, Solna
20   *World Paper* (1993) vol 218 (12), Dec, London
21   Women's Environmental Network (1990) *A Tissue of Lies? Disposable Paper and the Environment* WEN, London
22   Figures calculated from *Pulp and Paper International* (1988) quoted in Kroesa, R (1990) *op cit* Ref 2
23   Greenpeace (1990) *No Time To Waste* broadsheet produced by Greenpeace Canada, Vancouver
24   Kroesa, R (1990) *op cit* Ref 2
25   International Labour Organization (1992) *Social and Labour Issues in the Pulp and Paper Industry* ILO, Geneva
26   Stolton, S and N Dudley (1994) *The Timber Trade in Russia: A Report to WWF UK* Equilibrium Consultants, Bristol and Machynlleth
27   Worldwatch Institute, Washington DC
28   Karlsson, K (1994) personal communication to WWF UK 95 Group in Finland
29   Women's Environmental Network (1990) *op cit* Ref 21; and *Paper* (1992) 217 (2), Feb 1992
30   Kondo, T (1993) 'Japan's Pulp and Paper Industry Faces Dramatic Changes' *Paper & Packaging Analyst* no 15, Nov 1993, Pira International, UK
31   Women's Environmental Movement (1990) *op cit* Ref 21
32   Speranskaya, O (1993) 'Russia: Hope Again' *Paper* 128 (10), Oct 1993
33   Women's Environmental Network (1989) *The Sanitary Protection Scandal* WEN, London; and National Swedish Environmental Protection Board (1989) *Dioxins: A Programme for Research and Action* Stockholm, Sweden
34   Rathje, W and Murphy (1992), *Rubbish: The Archaeology of Garbage* New York

35   Figures taken from Pollock, C (1987) *Mining Urban Wastes: The Potential for Recycling* Worldwatch Paper 76, Worldwatch Institute, Washington DC; and Postell, S and J C Ryan (1991) 'Reforming Forestry' in *State of the World 1991* Worldwatch Institute, Washington DC

36   References from *Pulp and Paper International* 1987 and 1988; and World Resources Institute, Washington DC

37   References from *Conservatree ESP Magazine* USA; and Virtanen, Y and S Nilsson (1993) *Environmental Impacts of Waste Paper Recycling* International Institute for Applied Systems Analysis, Laxenburg, Austria

38   House of Lords Committee on the European Communities (1993) *Packaging and Packaging Waste* HMSO, London

39   FAO (1993) *FAO Yearbook: Forest Products 1980–1991* FAO Forestry Series no 26, FAO Statistics Series no 110, Food and Agriculture Organization of the United Nations, Rome

40   FAO (1993) *op cit* Ref 39

41   Thomas, B (1993) 'Germany's Paper Industry Postunification: From Boom to Recession' *Paper & Packaging Analyst* no 15, Nov 1993

42   Thies, C (1994) 'Increasing Demand for Clear-Cut Free Paper' *Taiga News* no 10, July 1994, Taiga Rescue Network, Jokkmokk, Sweden

43   Thomas, B (1993) *op cit* Ref 41

CHAPTER 7

1   Nectoux, F (1985) *Timber! An Investigation of the UK tropical timber industry* Friends of the Earth, London

2   Nectoux, F and N Dudley (1987) *A Hard Wood Story: Europe's Involvement in the Tropical Timber Trade* Friends of the Earth, London

3   See, for example, Counsell, S (1988) *The Good Wood Guide* Friends of the Earth, London

4   See, for example, Elliott, C and F Sullivan (1991) *Incentives and Sustainability – Where is ITTO Going?* WWF Position Paper, Gland

5   Buschbacher, R, C Elliott, D Reed and F Sullivan (1990) *Reforming the Tropical Forestry Action Plan: A WWF Position* WWF International, Gland, Sept 1990

6   Nectoux, F and Y Kuroda (1989) *Timber from the South Seas* WWF International, Gland

7   Lanley, J-P and J Clements (1978) *Present and Future Forest and Plantation Areas in the Tropics* FO/MISC/79/1 FAO, Rome

8   See, for example, Linear, M (1985) *Zapping the Third World, the Disaster of Development Aid* Pluto Press, London

9   FAO (1985) *The Tropical Forestry Action Plan* FAO in association with IBRD, WRI and UNDP, Rome

10   Colchester, M and L Lohmann (1990) *The Tropical Forestry Action Plan: What Progress?* World Rainforest Movement and *The Ecologist*, Penang and Sturminster Newton, Dorset

11   Jeanrenaud, J-P (1990) *An Assessment of the Conservation Significance of the Tropical Forestry Action Plan for Cameroon,* a report for WWF UK, Godalming

12   Buschbacher, R, C Elliott, D Reed and F Sullivan (1990) *op cit* Ref 5

13   Sargeant, C (1990) *Defining the Issues: Some Thoughts and Recommendations on Recent Critical Comments on TFAP* International Institute for Environment and Development, London

14   Winterbottom, R (1990) *Taking Stock: The Tropical Forestry Action Plan After Five Years* World Resources Institute, Washington DC

15   Commission on Transnational Corporations (1990) *Transnational Companies and the Environment,* report to the Economic and Social Council no E/C10/1990/10, 28 March 1990

16   Callister, D and S Broad (1995) *CITES and Trees: The Facts and the Fiction,* WWF-UK, Godalming

17   ITTO (1992) *ITTO Guidelines for the Sustainable Management of Natural Tropical Forests* ITTO
     Policy Development Series 1, ITTO, Yokohama, Japan; ITTO (1992) *Criteria for the
     Measurement of Sustainable Tropical Forest Management* Policy Development Series 3, ITTO,
     Yokohama; ITTO (1993) *ITTO Guidelines for the Establishment and Sustainable Management of
     Planted Tropical Forests* Policy Development Series 4, ITTO, Yokohama
18   Callister, D (1992) *Renegotiation of the International Tropical Timber Agreement: Issues Paper*
     TRAFFIC and WWF, Gland, Oct 1992
19   Elliott, C and F Sullivan (1991) *op cit* Ref 4
20   Callister, D (1992) *op cit* Ref 18
21   Hayter, T (1989) *Exploited Earth: Britain's Aid and the Environment: A Friends of the Earth
     Enquiry* Earthscan, London
22   Drawn from World Bank reports and listed in Nectoux, F and N Dudley (1987) *op cit* Ref
     2; Dudley, N (1986) *UK and International Aid and Tropical Forests,* report to Friends of the
     Earth, Earth Resources Research, London; Schwartzman, Stephen (1986) *Bankrolling
     Disaster: Development Banks and the Global Environment* Sierra Club USA; and Anon (undated)
     Financing Ecological Disaster: The World Bank and International Monetary Fund, a pamphlet
     endorsed by 28 environmental groups in 10 countries
23   World Bank (1990) *World Bank Forest Policy Brief* Washington DC; Buschbacher, R, C
     Elliott, D Reed and F Sullivan (1990) *The World Bank's New Forest Policy Brief: A WWF
     Position* WWF International, Gland, Switzerland
24   Anon (1994) 'European News' *World Paper* Sept 1994
25   Chatterjee, P (1994) *Fifty Years is Enough!* Friends of the Earth, London

CHAPTER 8

1   Rosamund, G (1994) *The Plantation Effect* Greenpeace New Zealand, Auckland
2   Fleet, H (1984) *New Zealand's Forests* Heinemann, Wellington
3   Halkett, J (1991) *The Native Forests of New Zealand* GP Publications Ltd, Auckland
4   Godley, E J (1975) 'Flora and Vegetation' in *Biogeography and Ecology in New Zealand* ed G
     Kuschel, Dr W Junke BV, Publishers, The Hague
5   Hackwell, K (unpublished), *Status Report on New Zealand's Temperate Forests* for the Sol III
     Temperate and Boreal Forests Project
6   See, for example, Holland, L (1989) *Wilderness Not Woodchips: Saving New Zealand's
     Rainforests* produced by the Royal Forest and Bird Protection Society for the New Zealand
     Rainforests Coalition; Jowitt, D (1991) *These Hills Are Tapu* Thames Coast Protection
     Society, Thames, New Zealand; Searle, G (1975) *Rush to Destruction* Friends of the Earth
     New Zealand, A H and A W Reed, Wellington; Williams, G R and D R Givens (1981)
     The Red Data Book of New Zealand Nature Conservation Council, Wellington
7   Anon (1989) A *National Policy for Indigenous Forests: A Discussion Paper* prepared by a Working
     Party Convened by the Secretary for the Environment, Ministry of the Environment,
     Wellington
8   Hackwell, K (unpublished) *op cit* Ref 5
9   New Zealand Forest Accord, signed Aug 1991
10   Juday, G P (1990) interview with J F Franklin, Bloedel Professor of Forestry, University of
     Washington, *Natural Areas Journal* **10** (4), pp 163–72
11   Gilmour, D A and R J Fisher (1991) *Villagers, Forests and Foresters: The Philosophy, Process and
     Practice of Community Forestry in Nepal* Sahayogi Press, Kathmandu, Nepal
12   Swanson, F J and J F Franklin (1991) *New Forestry Principles from Ecosystem Analysis of Pacific
     Northwest Forests,* paper given to AAAS meeting, Washington DC, 19 Feb 1991, submitted
     to *Ecological Applications*
13   Franklin, J E, D A Perry, T D Schowalter, M E Harmon, A McKee and T A Spies (1989)
     'Importance of Ecological Diversity in Maintaining Long-Term Site Productivity',

chapter 6 in *Maintaining the Long-Term Productivity of Pacific Northwest Forest Ecosystems* ed D A Perry, R Meurisse, B Thomas, R Miller, J Boyle, J Means, C R Perry and R F Powers, Timber Press, Portland, Oregon

14　Adams, P W, A L Flint and R L Fredriksen (1991) 'Long-Term Patterns in Soil Moisture and Revegetation after a Clearcut of a Douglas-Fir Forest in Oregon' *Forest Ecology and Management* **41**, pp 249–63

15　Spies, T A and S P Cline (1988) 'Coarse Woody Debris in Forests and Plantations in Coastal Oregon' in *From the Forest to the Sea: A Story of Fallen Trees* [ed] C Maser, R F Tarrant, J M Trappe and J F Franklin, USDA Forest Service, General Technical Report PNW-GTR-229, Portland, Oregon

16　Hansen, A J, T A Spies, F J Swanson and J L Ohmann (1991) 'Conserving Biodiversity in Managed Forests: Lessons from Natural Forests' *BioScience* **41** (6), pp 382–92

17　Franklin, J F and R T T Foreman (1987) 'Creating Landscape Patterns by Forest Cutting: Ecological Consequences and Principles' *Landscape Ecology* **1** (1), pp 5–18

18　Perry, D A (1988) 'Landscape Pattern and Forest Pests' *The Northwest Environmental Journal* **4**, pp 212–28, University of Washington, Seattle

19　Atkinson, W (1990) 'Another View of New Forestry', paper delivered at Annual Meeting, Oregon Society of American Foresters, Eugene, Oregon, 4 May 1990

20　DellaSala, Dominick A, David M Olsen, and Saundra L Crane (forthcoming) 'Ecosystem management and biodiversity conservation: applications to inland Pacific Northwest forests', to be published in *Proceedings of a Workshop on Ecosystem Management in Western Interior Forests* edited by D Baumgartner and R Everett, Washington State University Cooperative Extension Unit, Pullman, WA

21　MoDo (1991) Annual Report

22　Angelstam, P (1995) talk to field trip organized by WWF Sweden

23　Anon (1994) *Scandinavian Forestry: The forest should be used wisely – not used up*, Swedish Pulp and Paper Association, Svensk Skog, Norwegian Pulp and Paper Association, The Norwegian Forest Owners' Federation, Finnish Forest Industries Federation, Finnish Forestry Association, Nordic Timber Council

24　Arnqvist, R, Thorén, A and Ott, J (1994) *A Search for Sustainable Forestry – The Swedish View*, Skogsindustrierna, Stockholm

25　Three undated publications from Stora describe the changes: *Ecological Landscape Planning: The Grangärde Area as a model*, *White-Backed woodpecker* and *The Nature Conservation Strategy*, Stora Skog, Falun, Sweden

26　Gustafsson, J and Pettersson, B (1995) *Stora Forest Green Balance Sheet 1994*, Falun, Sweden

27　Juhlander, P (undated) *The New Forestry: Forestry – Nature's own way*, AssiDomän Skog & Trä, Falun; and Anon (undated) *Forestry's Green Revolution*, AssiDomän Skog & Trä, Falun

28　Lundberg & Co (1995) *The New Genesis: A book about the new forestry from Korsnäs*, Korsnäs, Gävle, Sweden; and Anon (1995) *Ecological Landscape Planning in the Vällen district*, Korsnäs, Gävle, Sweden

29　Information given on a WWF Sweden field trip in southern Sweden, May 1995

30　Section on RIL based on a presentation by S Yaacob, forestry officer at WWF Malaysia, at WWF offices, Kuala Lumpur, March 1994

CHAPTER 9

1　Read, M (1994) *Truth or Trickery? Timber Labelling Past and Future* WWF UK, Godalming

2　Adapted from Sullivan, F and J-P Jeanrenaud (1993) *The Inevitability of Timber Certification* WWF UK, Godalming

3　Some sections based on de Haes, C (1993) *WWF's Perspectives on the International Timber Trade and the Need for Credible Timber Certification*, speech at the first Euro-seminar on the timber trade and forest management, 'Promoting a Trade in Sustainably Produced

Timber' Brussels, 19 March 1993

4   Jeanrenaud, J-P and F Sullivan (1994) *Independent Certification: The Future for Forests,* paper presented at the *Financial Times* World Pulp and Paper Conference, London, 17–18 May 1994

5   Forest Stewardship Council (1995) *Fact Sheet* Oaxaca, Mexico

6   Adams, P (1992) 'Sustainable Forestry' *Living Earth: The Journal of the Soil Association* no 177, Soil Association, Bristol; documents from the Soil Association's Sustainable Forestry programme, and press release for the launch of the Woodmark standards

7   Jeanrenaud, J-P and F Sullivan (1994) *Timber Certification and the Forest Stewardship Council: A WWF Perspective* WWF UK, Godalming

8   Elliott, C and C A Clarke (1995) Memorandum from WWF International, Gland, Switzerland, 11 July 1995

9   WWF UK (1995) 'Progress Over the Last Six Months' *Forest Newsletter* **5**, Godalming

10   1995 Group Retailers Position Statement (1995) *Wood procurement Statement,* Jan 1995

11   Dudley, N and J-P Jeanrenaud (1994) Finland Study Tour by Members of the WWF UK 1995 Group to Research Options for Timber Certification, WWF UK and Equilibrium, Godalming, Surrey

12   Read, M (1994) *op cit* Ref 1

CHAPTER 10

1   WWF International (1994) *A Draft Global Forest Strategy* WWF International, Gland

2   UK Forest Network (in press) *A Forest Memorandum* UKFN, Norwich

3   Read, M (1994) *Truth or Trickery? Timber Labelling Past and Future* WWF UK, Godalming

Index

Also from Earthscan
The Forest Certification Handbook
by Christopher Upton and Stephen Bass

'The new Intergovernmental Panel on Forests will be addressing the transition to sustainable forest management. One of the key issues for its deliberation is the possible scope for forest certification as a means for opening up markets to the products of sustainably-managed forests. It should be a very helpful input into the decisions that need to be made at the international level, as well as within individual nations and enterprises'

Sir Martin Holdgate, former Director General, IUCN

'This is a clear and balanced overview of the key issues relating to timber certification. I strongly recommend it'

Chris Elliot, Senior Forest Officer, WWF International

'Explains clearly how a certification programme should be run, discusses critically what certification may or may not achieve in terms of solving the problems that beset forests, and gives an up to date account of the relationship between certification and other efforts that are being made to improve forest management worldwide. I strongly recommend it to readers, whether they be forest managers, traders, government officials or simply those who are concerned for the future of forests'

from the Foreword by Duncan Poore

From forester to retailer, stakeholders in the industry are under increasing pressure to assure customers that their wood products have come from well managed, sustainable forests. *The Forest Certification Handbook* gives practical advice on developing, selecting and operating a certification programme which provides both market security and raises standards of forestry management. It provides a thorough analysis of all the issues surrounding certification, including the commercial benefits to be gained, the policy mechanisms required, the interpretation and implementation of forestry management standards, and the process of certification itself. Three unique directories give details of currently certified forests, international and national initiatives, and active certification programmes.

Paperback £19.95 192pp ISBN 185383 222 7